S0-AUQ-167

FIBRE BUNDLES:
THEIR USE IN PHYSICS

FIBRE BUNDLES: THEIR USE IN PHYSICS

International Centre for Theoretical Physics,
Trieste, Italy
27 April – 1 May 1987

Editors
J P EZIN
A VERJOVSKY

Published by

World Scientific Publishing Co. Pte. Ltd.
P.O. Box 128, Farrer Road, Singapore 9128

U.S.A. office: World Scientific Publishing Co., Inc.
687 Hartwell Street, Teaneck NJ 07666, USA

FIBRE BUNDLES : THEIR USE IN PHYSICS

ISBN 9971-50-644-0

Printed in Singapore by Utopia Press.

AVANT–PROPOS

Les textes de ce recueil sont les exposés du Séminaire: "Espaces fibrés: leur utilisation en physique", tenu au Centre International de Physique Théorique de Trieste en Italie du 27 avril au 1er mai 1987.

Il s'agissait de présenter à un public de mathématiciens et de physiciens des exemples courants d'utilisation de la théorie des espaces fibrés en physique et en géometrie riemannienne.

Les sujets concernés recouvrent certains aspects de la géometrie des espaces des spineurs et des twisteurs, des théories de jauge et des monopoles, des opérateurs différentiels et pseudo–différentiels, des équations d'Einstein et d'Hamilton.

En publiant ces notes, en dépit des limites techniques qui ont contraint les auteurs à des exposés succincts, nous espérons mettre à la disposition du lecteur dans un même recueil des informations disséminées dans une vaste litterature. Il pourra approfondir les sujets abordés grâce aux indications et références données par les auteurs dans leurs exposés.

Nous remercions le Professeur James Eells, Directeur de la Section Mathématiques du Centre International de Physique Théorique d'avoir pris l'initiative d'organiser ce Séminaire et le Professeur Abdus Salam, Directeur de ce Centre d'avoir fourni les moyens financiers et l'hospitalité.

Les editeurs
Jean-Pierre EZIN
Alberto VERJOVSKY
International Centre for Theoretical Physics
34100 Trieste (Italy)

REMERCIEMENTS

Les éditeurs de ce recueil voudraient exprimer leurs remerciements au Centre International de Physique Théorique pour avoir financé cette réunion et pour avoir mis à disposition des participants leurs facilités de rencontre.

Ils remercient en outre la Section de Mathématiques de ce Centre pour l'organisation efficace du séminaire et le Service des Publications pour les soins et la rapidité apportés à la mise en page de ce livre.

TABLE DES MATIERES

FIBRE BUNDLES:
THEIR USE IN PHYSICS

I: Sur la géometrie des espaces des spineurs et des twisteurs

TRANSFORMATIONS OF SPINORS AND SPIN STRUCTURES UNDER DIFFEOMORPHISMS

Ludwik DABROWSKI

Scuola Internazionale Superiore di Studi Avanzati,
Trieste, Italy

and

Institute of Theoretical Physics, University of Wroclaw, Poland.

ABSTRACT

A pedagogical discussion of the transformation of spinor fields and spin structures on the example of two-dimensional torus is presented.

One way to describe a tensor field t on a manifold M is to specify its components with respect to reference frames. Namely, t can be thought of as an equivariant function from the bundle of frames LM over M into some linear space V, i.e. $t : LM \to V$ satisfies

$$t(eh) = D(h^{-1})\, t(e)$$

where eh denotes the (right) action of $h \in GL(n)$ on a frame $e = (e_1,\ldots,e_n) \in LM$ and $D: GL(n) \to L(V)$ is a representation of GL(n) in V. If one has a metric and an orientation on M then LM can be reduced to the bundle of oriented orthonormal frames F and GL(n) to the special orthogonal group SO(n). (For simplicity we restrict ourselves to the riemannian case, this can be generalized to metrics of any signature and nonorientable spaces.)

For spinor fields ψ one essentially repeats this construction i.e. if ρ: Spin(n) $\to$ SO(n) is the (nontrivial) double covering and D: Spin(n) $\to$ L(S) is a representation in S then

$$\psi: \tilde{F} \to S,$$
$$\psi(eh) = D(h^{-1})\, \psi(e)$$

for $e \in \tilde{F}$, $h \in$ Spin(n). Here $\tilde{F}$ is a principal Spin(n)-bundle, and, to be consistent, there should also be given a spin structure, i.e. a bundle morphism

$$\eta : \tilde{F} \to F,$$

such that

$$\eta(eh) = \eta(e)\,\rho(h). \tag{1}$$

It is known that M is orientable iff $w_1 = 0$ and a spin structure exists iff $w_2 = 0$, where w_i, i=1,2 are the Stiefel-Whitney classes of M. Also, inequivalent spin structures are labelled by the first cohomology group $H^1(M,Z_2)$ of M with coefficients in Z_2, where $\eta: \tilde{F} \to F$ and $\eta': \tilde{F}' \to F$ are equivalent iff there exists a bundle isomorphism $\beta: \tilde{F}' \to \tilde{F}$ such that

$$\eta \circ \beta = \eta'. \tag{2}$$

These purely topological properties and alternative definitions of the spin structure [M] show that the role of a metric g is not very essential. One might investigate, for instance, the general covariance of a model with spinors and the transformation of spinors under diffeomorphisms which are not isometries of g and do not preserve F. A problem to do this is that by requiring dim $S < \infty$, one cannot use any spinor representation of $\widetilde{GL}_0(n)$ (the nontrivial double cover of $GL_0(n)$). To the best of our knowledge an intrinsic definition of the action of the group Diff(M) of orientation preserving diffeomorphisms on spinors and on spin structures was first written down in [D & P1]* and shown to agree with the transformation rule for local components of ψ used in General Relativity. As an intermediate step it makes use of a prolongation $\alpha \equiv [\widetilde{LM}, \eta]$, of the bundle of all oriented frames LM to the structure group $\widetilde{GL}_0(n)$, where $\eta: \widetilde{LM} \to LM$ has the property analogous to (1). One asks for the possibility of lifting Tf: LM $\to$ LM (the derivative of $f \in$ Diff(M)) to a bundle morphism $\widetilde{Tf}: \widetilde{LM}' \to \widetilde{LM}$ which intertwines (possibly inequivalent) prolongations α' and α

----

* Prof. Cahen informed me that a similar construction appeared in a lecture by J.P. Bourguignon.

$$\eta \circ \widetilde{Tf} = Tf \circ \eta'. \qquad (3)$$

(Note that (3) reduces to (2) if $f = id_M$.) It turns out that for each f and each α such α' always exists and is unique. This defines the transformation of prolongations (and, equivalently, of spin structures), denoted by $\alpha \to p_f(\alpha)$. Moreover $p_{fof'} = p_f \circ p_{f'}$ and we have an action of Diff(M), which passes to the quotient $\Omega(M) = $ =Diff(M)/$Diff_0$(M)since $p_f = p_{f'}$ if f and f' are isotopic.

Note that $f \in$ Diff(M) can only mix two prolongations which have equivalent bundles. Next, for each $x \in M$ there are precisely two morphisms from $\widetilde{LM}_x$ the fibre over x to the fibre $\widetilde{LM}_{f(x)}$ over f(x), which differ by $-1 \in$ Spin(n) $\in GL_0(n)$, and both satisfy eq. (3). Starting from some $x_0 \in M$, one can choose one of them $\widetilde{Tf}_x$ by continuity along a path from x_0 to x. Then, $\widetilde{Tf}$ exists globally iff different paths give the same answer. This, in turn, means that the $\pi_1(M) \to Z_2$ group homomorphism which we obtain in this way should be trivial. Since $H_1(M,Z)$ is an abeliazation of $\pi_1(M)$, it is sufficient (and obviously necessary) to check the equation (3) over the loops generating $H_1(M, Z)$ in order to determine p_f.

Now, by composing ψ with $\widetilde{Tf}$ one gets the transformed spinor field $\psi' = \psi o \widetilde{Tf}$ which, strictly speaking, is associated with the metric f*g and with the spin structure $p_f(\alpha)$. An important issue is that there always exist exactly two lifts $\widetilde{Tf}$ and $\gamma o \widetilde{TF}$ of Tf between $p_f(\alpha)$ and α, where γ is the rigid multiplication by -1. Therefore, restricting our attention to $Diff_0$(M) which preserves all the spin structures, it is rather the double covering D^α of $\mathrm{Diff}_0(M)$ which acts on spinors. Here $D^\alpha=\{u \in \mathrm{Aut}\,\widetilde{LM} \mid \eta \circ u = Tf \circ \eta$ for some $f \in \mathrm{Diff}_0(M)\}$, and $\kappa^\alpha : D^\alpha \to \mathrm{Diff}_0(M)$ is given by $\kappa^\alpha : u \mapsto f$. Note, that D^α and κ^α may depend on α.

We illustrate these general properties by the example of a two-dimensional torus $T = S^1 \times S^1$ with mod 2π periodic coordinates (x, y). Trivialize $LM = T \times GL_0(2)$ by the global two-frame $(\partial/\partial_x, \partial/\partial_y)$. Since $H^1(T, Z_2) = \mathrm{Hom}(\pi_1(T), Z_2)$ and $\pi_1(T) = Z \times Z$, there are four spin structures on T which we label $\alpha_{j,k}$ with j,k=0 or 1. They are defined by $\widetilde{LM} = T \times \widetilde{GL}_0(2)$ together with bundle morphisms

$$\eta_{j,k} : (x,y,h) \mapsto (x,y,\rho(h)\ R(jx + ky)), \tag{4}$$

where

$$R(\alpha) \equiv \begin{pmatrix} \cos\alpha & \sin\alpha \\ -\sin\alpha & \cos\alpha \end{pmatrix} \in SO(2).$$

To see that $\alpha_{j,k}$ and $\alpha_{j',k'}$ are equivalent iff $j = j'$ and $k = k'$ observe that a bundle automorphism β must necessarily be of the form

$$\beta(x,y,h) = (x,y,\hat{\beta}(x,y)\, h),$$

where $\hat{\beta} : T \to \widetilde{GL}_0(2)$ is smooth. Then, the condition (2) means that

$$R(j'x + k'y) = \rho(\hat{\beta}(x,y))\, R(jx + ky)$$

i.e. $\hat{\beta}(x,y)$ should be a smooth lift of $R((j'-j)x+(k'-k)y)$ (c.f.[I]). Checking this condition separately along the loops $a(s) = (s,0) \in T$ and $b(s) = (0,s) \in T$ yields that $j=j'$ and $k=k'$

(ρ is the nontrivial double covering!) Next, we apply this method to find the action of $\Omega(T)$ on $\alpha_{j,k}$. The group $\Omega(T)$ is generated by (the class of) two Dehn twists

$$f_a: (x,y) \mapsto (x+y, y)$$

$$f_b: (x,y) \mapsto (x, x+y).$$

Now, $Tf_a: (x,y,h) \mapsto (x+y, y, \begin{pmatrix} 1 & 1 \\ 0 & 1 \end{pmatrix} h)$ and the most general form of $\widetilde{Tf}_a$ is

$$\widetilde{Tf}_a: (x,y,h) \to (x+y, y, \widehat{Tf}_a(x,y)h),$$

where $\widehat{Tf}_a : T \mapsto \widetilde{GL}_o(2)$ is smooth. From eq.(3) it follows that

$$R(j(x+y) + ky)\rho(\widehat{Tf}_a(x,y)) = \begin{pmatrix} 1 & 1 \\ 0 & 1 \end{pmatrix} R\,(j'x + k'y)$$

i.e. $\widehat{Tf}_a(x,y)$ must be a smooth lift of

$$R(-j(x+y)-ky) \begin{pmatrix} 1 & 1 \\ 0 & 1 \end{pmatrix} R\,(j'x + k'y).$$

By observing that the topology of the covering $\widetilde{GL}_o(2) \to GL_o(2)$ is totally contained in $Spin(2) \to SO(2)$ (these are maximal compact subgroups) we may contract $\begin{pmatrix} 1 & 1 \\ 0 & 1 \end{pmatrix}$ to $\begin{pmatrix} 1 & 0 \\ 0 & 1 \end{pmatrix}$. Then, checking this condition along a(s) and b(s) we get $j = j'$ and $k' = j+k$, i.e. $p_{fa} : \alpha_{j,k} \mapsto \alpha_{j,\, j+k}$. By a similar consideration we obtain $p_{fb} : \alpha_{j,k} \mapsto \alpha_{j+k,\, k}$.

Next we shall find the coverings $\kappa^{j,k}: D^{j,k} \to Diff_o(T)$. Since the maximal compact subgroup of $Diff_o(T)$ is the group $U(1) \times U(1)$ of 'translations' [E&E] we shall discuss only the transformations

$$f_{s,t}: (x,y) \mapsto (x+t, y+s),$$

where $s,t \in (0, 2\pi)$. We have $Tf_{s,t}: (x,y,h) \to (x+t, y+t, h)$ and $\widetilde{Tf}_{s,t}$ must be of the form $\widetilde{Tf}_{s,t} : (x,y,h) \to (x+t, y+t, \widehat{Tf}_{s,t}(x,y)h)$. From eq.(3) it follows that $\widehat{Tf}_{s,t}(x,y)$ should be a lift of $R(-j(x+t)-k(y+s))\, R(jx+ky) = R(-jt-ks)$. If we now fix x,y and s, and vary smoothly the parameter t from 0 to 2π we get that $\widehat{Tf}_{s,t}$ jumps by $(-1)^j$. Similarly, if we vary s then the jump is $(-1)^k$. Therefore, we get the following four distinct double coverings of $U(1) \times U(1)$:

(j,k)	$(U(1) \times U(1))^{j,k}$	$\kappa^{j,k}$
(0,0)	$U(1) \times U(1) \times Z_2$	$(t,s, \pm 1) \mapsto (t,s)$
(1,0)	$U(1) \times U(1)$	$(t,s) \mapsto (2t,s)$
(0,1)	$U(1) \times U(1)$	$(t,s) \mapsto (t, 2s)$
(1,1)	$(U(1)\times U(1))/\{(1,1),(-1,-1)\}$	$[t,s] \mapsto (2t, 2s)$.

In the terminology of [A&H] the action of the first (respectively second) U(1) factor is of even type if (j,k) = (0,0) or (0,1) (respectively (0,0) or (1,0)) and otherwise is of odd type.

We conclude with some remarks. In the particular case of complex Kahler manifolds (e.g. $T = S^1 \times S^1$) one may simplify the general discussion in terms of square roots of the canonical holomorphic bundle [Hit]. Also, to investigate the action of compact groups on spin structures on riemannian manifolds one could use for instance an invariant metric, but the method in [D&P1] is most general (e.g. lorentzian metrics) and has been applied to all compact two-dimensional manifolds [D&P2]. Despite the fact that we can transform a spinor field under an arbitrary diffeomorphism, the problem to define intrinsically its Lie derivative is still open (if spinors appear only in scalar combinations a possibility is to use the covariant derivative).

References

[A] Atiyah, M.F.: "Spin manifolds and group actions in Essays on Topology and Related Topics", pp. 18-28, Springer Verlag, New York, 1970.

[D&P1] Dąbrowski, L. and Percacci, R.: "Spinors and Diffeomorphisms", Commun.Math.Phys., 106, 691, 1986.

[D&P2] Dąbrowski, L. and Percacci, R.: "Diffeomorphisms, orientation and pin structures in two dimensions", SISSA preprint 103/86/E.P.

[E&E) Earle, C.J. and Eells, J.: "The diffeomorphism group of a compact Riemann surface", Bull.Ann.Math.Soc., 73, 557, 1967.

[H] Hitchin N.:Harmonic Spinors, Adv. Math. 14, 1, 1974.

[I] Isham, C.J.: "Spinor fields in four dimensional space-time", Proc.R.Soc., London, A364, 591 (1978).

[M] Milnor, J.: "Spin structures on manifolds", L'enseignement Math., 9, 198 (1963).

HARMONIC MAPS, TWISTOR SPACES AND TWISTOR LIFTS

Gary R. Jensen

Department of Mathematics, Washington University,
St. Louis, Missouri 63130, USA

and

Marco Rigoli

International Centre for Theoretical Physics, Trieste, Italy.

ABSTRACT

This paper describes the geometry of the twistor space of a 4–manifold and also the relation of twistor lifts to study harmonic maps from Riemann surfaces into four manifolds.

1. INTRODUCTION

Let (M,g) and (N,h) be two Riemannian manifolds, M compact. Given a smooth map $f: M \to N$ one defines the energy of f to be

$$E(f) = \frac{1}{2}\int_M |df|^2 dV_g$$

where $|\quad|$ is the Hilbert–Schmidt norm of df considered as a section of the bundle $TM^* \otimes TN$ and dV_g is the volume element of the metric g. Considering E as a functional on the (infinite dimensional) manifold $C^\infty(M,N)$ of smooth maps from M to N one can define critical points of E. Such critical points are called harmonic maps and they satisfy the Euler–Lagrange equation

$$\tau(f) = 0 \tag{1.1}$$

where the tension field τ of the map f is defined to be

$$\tau(f) = tr_g \nabla df$$

for ∇ the natural connection induced on $TM^* \otimes TN$. In case $N = \mathbf{R}^n$, with its canonical flat structure, (1.1) reduces to $\Delta_g f = 0$ where Δ_g is the Laplace–Beltrami operator of the metric g and we can therefore think of (1.1) as a natural generalization of the wave equation to scalar fields required to obey some non linear constrain. A typical example is given by three independent fields φ^k , $k = 1,2,3$ parametrizing four meson fields π^k, σ that satisfy $\pi^k\pi^k + \sigma^2 =$ const. as in the non linear σ–model proposed by Gell–Mann and Levy [1] in 1960.

Before proceeding any further observe that (1.1) makes sense even if M is non compact and that if f is an isometry, that is, $f^*k = g$, (1.1) reduces to

$$H = 0 \tag{1.2}$$

where H is the mean curvature vector of the immersion. In fact (1.2) is satisfied iff f is a critical point of the volume functional. For details and as a general reference in the theory of harmonic maps we refer the reader to the paper of Eells and Lemaire [2]. In the case M is a Riemann surface the notion of harmonic map is independent of the representative metric in the conformal class defining the complex structure of M and one verifies that for M compact and N a Kähler manifold a $\pm$ holomorphic map $f: M \to N$ minimizes the energy functional and is therefore a harmonic map. The natural problem is then to decide whether other critical points exist in the above situation and if eventually it is possible to express them in terms of holomorphic data. As shown in [3],[4],[5] this is in fact the case when $M = S^2$ and $N = \mathbf{C}P^n$ and these, in Physics jargon, σ–models have attracted attention as test cases for non abelian gauge theories. In fact conformal invariance in a particular dimension, existence of a topological charge number ([3]), the description of complicated solutions from simpler ones, are features common with the celebrated Yang–Mills theory. Even more, Atiyah [6] has recently described an important link between instanton solutions of Yang–Mills equations over S^4 and σ–models over S^2 *). In this paper we will describe some relations between harmonic maps from Riemann surfaces into a 4–dimensional target N and complex geometry, the underlying philosophy being to try to describe solutions of the quasi–linear elliptic system described by (1.1) (see [2] for an expression in local coordinates) in terms of holomorphic data on the Riemann surface M. The most famous example is of course the Eisenhart "holomorphic" representation of minimal surfaces in $\mathbf{R}^4$, and, since it will serve as a guideline to our analysis, let us briefly describe some of the crucial results in this setting. Given an immersed surface $f : M \to \mathbf{R}^4$ let $G_2(\mathbf{R}^4)$ be the Grassmann manifold of two planes in $\mathbf{R}^4$, then the Gauss map $\gamma_f : M \to G_2(\mathbf{R}^4)$ is defined by assigning to a point $p \in M$ the 2–dimensional plane $f_{*p}(T_pM)$. A special feature of $G_2(\mathbf{R}^4)$ is that it can be canonically identified with the projective quadric $Q_2 = \{[z] \in \mathbf{C}P^3 : z^2 = 0\}$ of isotopic directions in $\mathbf{C}^4$. With respect to this Kähler structure on $G_2(\mathbf{R}^2)$ a basic result of Chern [7] states that f is minimal if and only if γ_f is -holomorphic. Furthermore, peculiar to this quadric is the biholomorphic (isometric under a proper normalization) splitting $G_2(\mathbf{R}^4) \simeq \mathbf{C}P^1 \times \mathbf{C}P^1$ which allows us to define the two projections $\gamma_f^{\pm}$ of γ_f on the two factors. As can be deduced from the work of Hoffman–Osserman [10] the Eisenhart integral representation of f is obtained via $\gamma_f^{\pm}$ and an appropriate Abelian differential ν in the form

$$ {}^t f(p) = (f^1(p), \ldots, f^4(p)) = Re \int_\gamma \frac{1}{2} (1 + \gamma_f^+ \gamma_f^-, i(1 - \gamma_f^+ \gamma_f^-), \gamma_f^+ \gamma_f^-, -i(\gamma_f^+ + \gamma_f^-))\nu \quad (1.3) $$

γ is any curve connecting the fixed point $0 \in M$ to the generic point p. The fundamental role of $\gamma_f^{\pm}$ is revealed also in the non minimal case. It is indeed sufficient to recall the following two striking formulas due to Blaschke [8] and Hoffman and Osserman [9].

$$ K = J(\gamma_f^+) + J(\gamma_f^-), \quad K^{\perp} = J(\gamma_f^+) - J(\gamma_f^-) \quad (1.4) $$

where $J(\ )$ is the Jacobian of a map and $K, K^{\perp}$ are respectively the Gaussian and normal curvatures of the immersion. As a consequence, for M compact, using the Chern–Gauss–Bonnet theorem we have the topological results, generalizing Chern–Spanier (see [9]),

$$ \chi(M) = \deg(\gamma_f^+) + \deg(\gamma_f^-) \quad (1.5) $$

$$ \chi(TM^{\perp}) = \deg(\gamma_f^+) - \deg(\gamma_f^-) \quad (1.6) $$

where $\chi(\ )$ is the Euler characteristic and deg( ) is the degree. For a general 4–dimensional target N the role of $G_2(\mathbf{R}^4), \gamma_f, \mathbf{C}P^1$, $\gamma_f^{\pm}$ will be substituted by the Grassmann bundle

*) Further interesting relations can be found in the lectures of C.M. Wood in this same volume.

of 2–planes over N, the Gauss lift γ, the twistor space W over N and the twistor lifts $\varphi_\pm$ as we are going to show in the next sections. More precisely, Sec. 2 briefly deals with the Riemannian and $J_\pm$ Hermitian geometry of W, while Sec. 3 is devoted to the description of the Gauss and twistor lifts, the study of their holomorphicity, harmonicity properties and interrelations. The new material presented here is taken from a more comprehensive work of the authors and we refer to [11] for a complete and detailed treatment of the arguments we are going to introduce. For the rest of the paper M is a connected Riemannian surface with prescribed metric g in its conformal class and N a 4–dimensional oriented Riemannian manifold. We fix the index ranges $1 \leq A, B, C, \ldots \leq 4$, $1 \leq i, j, k, \ldots \leq 2$, $3 \leq \alpha, \beta, \gamma, \ldots \leq 4$ and use Einstein summation convention.

2. RIEMANNIAN AND HERMITIAN GEOMETRY OF THE TWISTOR SPACE

Let W be the set of pairs (p,J) where $p \in N$ and J is an orthogonal transformation of T_pM satisfying $J^2 = -id$ that is an orthogonal complex structure of T_pN. (Warning: we do not assume J to be orientation preserving). W is called the twistor space of N and the twistor projection $T : W \to N$ defined by

$$T : (p, J) \to p \tag{2.1}$$

makes W a fibre bundle over N with standard fibre $O(4)/U(2)$. We describe the Riemannian structure of W as follows: let $\pi : 0(N) \to N$ be the 0(4)–bundle of orthonormal frames of N and define a map $\sigma : 0(N) \to W$ by setting

$$\sigma : (p, e) \to (p, J_e) \tag{2.2}$$

where we have described points in $0(N)$ as pairs (p, e), $p \in N$, e an orthonormal basis of T_pN and J_e is the complex structure of T_pN determined by the requirements

$$J_e e_1 = e_2, \quad J_e e_3 = e_4, \quad J_e^2 = -id \tag{2.3}$$

σ satisfies $\sigma : (p, e)A \to (p, J_{eA})$ for $A \in 0(4)$ and it induces the bundle isomorphism

$$W \simeq O(N)/U(2) \tag{2.4}$$

Let $\theta = \{\theta^A\}$, $\omega = \{\omega^A_B\}$ denote the canonical and Levi–Civita connection forms respectively on $0(N)$. θ and ω satisfy the structure equations

$$d\theta = -\omega \wedge \theta, \quad d\omega = -\omega \wedge \omega + \Omega \tag{2.5}$$

for Ω the curvature form of components

$$\Omega^A_B = \frac{1}{2} R^A_{BCD}\, \theta^C \wedge \theta^D \tag{2.6}$$

Let A be the matrix

$$\begin{pmatrix} \begin{matrix} 0 & -1 \\ 1 & 0 \end{matrix} & 0 \\ 0 & \begin{matrix} 0 & -1 \\ 1 & 0 \end{matrix} \end{pmatrix}$$

and define for a non zero constant t

$$\nu_t = \frac{t}{2}\,(\omega + A\omega A) \tag{2.7}$$

Then the fibres of σ are the integral submanifolds of the completely integrable Pfaffian system

$$\theta = 0, \quad \nu_t = 0 \tag{2.8}$$

and the Riemannian metric ds_t^2 on W is characterized by $\sigma^* ds_t^2 = Q$ where Q is the U(2)–invariant symmetric bilinear form on 0(N)

$$Q = {}^t\theta\theta + < \nu_t, \nu_t > \tag{2.9}$$

for $<,>$ the bi–invariant metric $< x, y >= tr^t xy, \quad x, y \in \sigma(4)$ the Lie algebra of $0(4)$. In terms of components if $u : U \subset W \to 0(N)$ is a local section then an orthonormal coframe for ds_t^2 on U is given by the pull–back by u of $\varphi^A, \varphi^5, \varphi^6$ with

$$\varphi^A \equiv \theta^A, \quad \varphi^5 = t(\omega_3^1 - \omega_4^2), \quad \varphi^6 = t(\omega_4^1 + \omega_3^2) \tag{2.10}$$

The corresponding Levi–Civita connection forms φ_B^A are easily found to be

$$\begin{cases} \varphi_B^A = \omega_B^A + \frac{t}{2}(R_{13BA} - R_{24BA})\varphi^5 + \frac{t}{2}(R_{14BA} + R_{23BA})\varphi^6 \\ \varphi_B^5 = \frac{t}{2}(R_{24AB} - R_{13AB})\varphi^A = -\varphi_5^B \\ \varphi_B^6 = -\frac{t}{2}(R_{14AB} + R_{23AB})\varphi^A = -\varphi_6^B \\ \varphi_6^5 = \omega_2^1 + \omega_4^3 = -\varphi_5^6 \end{cases} \tag{2.11}$$

Computation of the curvature tensor is a standard one. In fact an investigation of the Ricci curvature gives the following

Theorem 2.1

Let (W, ds_t^2) be the twistor space of the 4–dimensional Einstein manifold N with scalar curvature R. Then for $R \leq 0$, (W, ds_t^2) can never be rendered Einstein for any choice of t, while for $R > 0$ the two choices $t^2 = \frac{6}{R}, \frac{12}{R}$ render (W, ds_t^2) Einstein.

We define two almost complex structures $J_\pm$ on W relevant to our purposes by defining a basis for the corresponding (1,0) forms. For J_+ we consider the basis

$$\rho^1 = \varphi^1 + i\varphi^2, \quad \rho^2 = \varphi^3 + i\varphi^3, \quad \rho^3 = \varphi^5 + i\varphi^6 \tag{2.12}$$

while for J_-, we take the basis $\rho^i, \overline{\rho}^3$. J_+ is the complex structure introduced by Atiyah–Hitchin–Singer [12] in their study of self dual Yang–Mills equations in Euclidean 4–space, while J_- has been studied by Eells–Salamon [13] in connection with harmonic maps. J_- turns out to be never integrable while integrability of J_+ is expressed in terms of the splitting of the Weyl tensor $\mathcal{W}$ of N into $\mathcal{W} = W^+ + W^-$ (Singer–Thorpe [14]). Recalling that N is said to be $\pm$ selfdual if and only if $W^\mp = 0$ the integrability property of J_+ is equivalent to selfduality of N [12]. In our setting this can easily be verified since integrability of J_+ is equivalent to the closedness of the differential ideal generated by ρ^1, ρ^2, ρ^3. Differentiation of these forms gives the necessary and sufficient condition

$$\Omega^1_3 - \Omega^2_4 + i(\Omega^1_4 + \Omega^2_3) \equiv 0 \quad \mathrm{mod}(\rho^1, \rho^2)$$

whose analysis in terms of the curvature tensor we leave to the reader [11]. It is worth noticing that W is invariant under a conformal change of the metric in N and J_+ is invariant too while J_- is not. This can in fact be proved in an elementary fashion by the transformation undertaken by ρ^1, ρ^2, ρ^3 under a conformal change of the metric h in N. (For another proof see Salamon [15], Theorem 12.1.)

The Kähler forms $k_\pm$ of $J_\pm$ are given by

$$k_\pm = \frac{i}{2}\left(\rho^1 \wedge \overline{\rho}^1 + \rho^2 \wedge \overline{\rho}^2 \pm \rho^3 \wedge \overline{\rho}^3\right) \tag{2.13}$$

and exterior differentiation of (2.13) gives the necessary and sufficient conditions in terms of the components of the curvature tensor of N for $J_\pm$ to be symplectic. Analysis of these components together with those of W^+ gives the following result first proved by Friedrich and Kurke [16]:

Theorem 2.2

Let (W, ds_t^2, J_+) be the twistor space of the 4 manifold N with scalar curvature R. Then $(W ds_t^2, J_+)$ is Kähler (that is J_+ is integrable and symplectic i.e. $dk_+ = 0$) if and only if N is $-$ self dual Einstein and $t^2 = \frac{12}{R}$.

Remark

From Theorem 2.1 (W, ds_t^2, J_+) is Einstein in the above assumptions. On the other hand [11]

Theorem 2.3

Let (W, ds_t^2, J_+) be the twistor space of the self-dual Einstein manifold N of negative scalar curvature R. Then for $t^2 = -\frac{12}{R}$, (W, ds_t^2, J_-) is symplectic, i.e. $dk_- \equiv 0$.

The above result gives a relatively simple example of a symplectic almost complex, not Kähler, Hermitian manifold, see also Thurston [17].

Recall that $J_\pm$ is called (1,2)–symplectic if $(dk_\pm)^{1,2} \equiv 0$ where with the notation $(\,)^{1,2}$ we mean the $(1,2)$ part. It is an easy matter to determine the necessary and sufficient conditions for (1,2)–symplecticity of $J_\pm$ in terms of the curvature of N, but what we are interested in here is only the relevance of such a notion with respect to harmonicity due to the following result first proved by Lichnerowicz [18].

Theorem 2.4

Let $f : M \to N$ be a holomorphic map from a Riemann surface to an almost Hermitian (1,2)–symplectic manifold. Then f is harmonic.

We remark that Eells–Salamon [13] give an example of holomorphic maps $f : M \to N$ which are harmonic even though N is not (1,2)–symplectic, so that the assumptions of the theorem are sufficient but not necessary, while on the other hand, an example of Gray reported in [2, section 9] shows that such assumptions cannot be weakened.

3. TWISTOR LIFTS AND THEIR GEOMETRIC ROLE

Let $\pi : G_2(TN) \to N$ be the Grassmann bundle of oriented 2–planes of N. Then a point of $G_2(TN)$ can be represented by a pair $(p, \{e_i\})$ with $p \in N$ and $\{e_i\}$ an oriented orthonormal basis of an oriented 2–plane in T_pN. We define two projections $\pi_\pm : G_2(TN) \to W$ as follows: for $(p, \{e_i\}) \in G_2(TN)$ complete $\{e_i\}$ to an oriented orthonormal basis $\{e_A\}$ of T_pN and define

$$\pi_\pm : (p, \{e_i\}) \to (p, \tilde{J}_\pm)$$

with $\tilde{J}_\pm$ defined by the requirements $\tilde{J}_\pm^2 = -id, \quad \tilde{J}_\pm e_1 = e_2, \quad \tilde{J}_\pm e_3 = \mp e_4$.

Given the isometric immersion $f : M \to N$ let $e = \{e_A\}$ be a Darboux frame along f, that is, $\{e_i\}$ span $f_{*p}(T_pM)$ at $f(p)$ with $e_1 \wedge e_2$ giving the correct orientation and $\{e_\alpha\}$ complete them in such a way that $e_1 \wedge e_2 \wedge e_3 \wedge e_4$ agrees with the orientation of $T_{f(p)}N$. Then the Gauss lift $\gamma : M \to G_2(TN)$ of f is defined by

$$\gamma : p \to (f(p), \{e_i\})$$

while the twistor lifts $\varphi_\pm : M \to W$ are defined by the composition

$$\varphi_\pm = \pi_\pm \circ \gamma.$$

Set

$$A_\pm = \begin{pmatrix} 1 & & & \\ & 1 & & \\ & & 1 & \\ & & & \mp 1 \end{pmatrix} \in 0(4)$$

then because of the definition of the $\varphi_\pm$ the frames $e_\pm = R_{A_\pm} \circ e$ satisfy

$$\sigma \circ e_\pm = \varphi_\pm \tag{3.1}$$

where $\sigma : 0(N) \to W$ has been defined in (2.2). This enables us to relate the geometry of f with that of $\varphi_\pm$ and show that the role the latter play is exactly analogue to the role played by the two projections on the $\mathbf{C}P^1$ factors described in the introduction for the case of a surface in $\mathbf{R}^4$. First let us describe the local geometry of the immersion f. Let θ be the coframe dual to the Darboux frame e. Then by the definition of e we have on M (for simplicity we will generally drop the pull–back notation)

$$\theta^\alpha = 0 \tag{3.2}$$

Differentiation of (3.2), use of the structure Eqs.(2.5) and Cartan's Lemma give

$$\theta_i^\alpha = h_{ij}^\alpha \theta^j \tag{3.3}$$

for some locally defined functions h_{ij}^α satisfying $h_{ij}^\alpha = h_{ji}^\alpha$. Such functions are the coefficients of the second fundamental tensor II with respect to the frame we are considering. The coefficients of the covariant derivative DII are given by the expression

$$h_{ijk}^\alpha \theta^k = dh_{ij}^\alpha - h_{kj}^\alpha \omega_i^k - h_{ik}^\alpha \omega_j^k + h_{ij}^\beta \omega_\beta^\alpha \tag{3.4}$$

where $h_{ijk}^\alpha = h_{jik}^\alpha$ while

$$h_{ijk}^\alpha = h_{ikj}^\alpha - R_{ijk}^\alpha \tag{3.5}$$

express the Codazzi equations. Letting H be the mean curvature vector then $H = 0$ if and only if h_{ii}^α while $DH = 0$ if and only if $h_{iik}^\alpha = 0$. In the above two cases f is respectively said to be minimal or with parallel mean curvature.

We will now consider (W, ds_t^2) with both J_+ and J_- structures at once. From (2.12) we obtain

$$\varphi_\pm^*(\rho^1) = \varphi, \quad \varphi_\pm^*(\rho^2) = 0 \quad \varphi_-^*(\rho^3) = -b\varphi - \overline{S}_-\overline{\varphi}, \quad \varphi_+^*(\rho^3) = -\overline{b}\varphi - \overline{S}_+\overline{\varphi} \tag{3.6}$$

while for J_- we have to substitute the last two equations of (3.6) with their complex conjugate. In (3.6) $\varphi = \theta^1 + i\theta^2$ is a basis for the (1,0) forms on M, while the coefficients $b, S_\pm$ are defined by

$$b = \frac{1}{2}\ h_{kk}^3 + \frac{i}{2}\ h_{kk}^4, \quad S_\pm = L^3 \pm iL^4, \quad L^\alpha = \frac{1}{2}\ (h_{11}^\alpha - h_{22}^\alpha) - ih_{12}^\alpha \tag{3.7}$$

Of course $|b| = |H|$ and if we set $s_\pm = |S_\pm|/\sqrt{2}$ we like to underline the remarkable fact that $s_\pm$ are contact invariants of f, thus, in particular, globally defined real functions on M (only $s_\pm^2$ are in general smooth) [11]. From (3.6) we derive the following conclusion due to Eells–Salamon [13] relevant to our present purposes (a detailed analysis can be found in [13] and different conclusions in [11])

$$\varphi_\pm \quad \textit{are} \quad J_- \quad \textit{holomorphic if and only if} \quad f \textit{ is minimal.} \tag{3.8}$$

At the end of Sec.2, we remarked on the result of Lichnerowicz, Theorem 2.4, that holomorphicity and harmonicity do not share a very strict dependence. It is therefore instructive to deal with the problem of harmonicity of $\varphi_\pm$ directly. To fix ideas (but the final result will be stated for φ_+ too) let us consider here φ_-. Let B^i, B^α, B^5, B^6 be the coefficients of the tension field of $\varphi_- : M \to (W, ds_t^2)$ then a computation performed by using the results introduced above gives

$$\begin{cases} B^i = t^2\{(R_{13ki} - R_{24ki}(h^4_{2k} - h^3_{1k}) - (R_{14ki} + R_{23ki})(h^4_{1k} + h^3_{2k})\} \\ B^\alpha = h^\alpha_{kk} + t^2\{(R_{13k\alpha} - R_{24k\alpha})(h^4_{2k} - h^3_{1k}) - (R_{14k\alpha} + R_{23k\alpha})(h^4_{1k} + h^3_{2k})\} \\ B^5 = t(h^4_{kk2} - h^3_{kk1} - R_{4k2k} + R_{3k1k}) \\ B^6 = -t(h^4_{kk1} + h^3_{kk2} - R_{4k1k} - R_{3k2k}) \end{cases} \tag{3.9}$$

If we assume that N is $-$ self dual and Einstein then an elementary analysis of the Einstein condition together with $W^+ = 0$ simplifies equations (3.9) to give the tension field of φ_-

$$\tau(\varphi_-) = h^\alpha_{kk}\left(1 - \frac{R}{12}t^2\right)E_\alpha + t(h^4_{kk2} - h^3_{kk1})E_5 - t(h^4_{kk1} + h^3_{kk2})E_6 \tag{3.10}$$

with R being the scalar curvature of N and $(E_i, E_\alpha, E_5, E_6)$ the dual frame to (2.10). We have therefore proved the following [11].

Theorem 3.1

Let $f : M \to N$ be an isometric immersion of a Riemann surface into a 4-dimensional -(+) selfdual Einstein manifold with scalar curvature R and consider the twistor lift $\varphi_-(\varphi_+) : M \to (W, ds_t^2)$.

i) If $R \leq 0$ then f is minimal if and only if $\varphi_-(\varphi_+)$ is harmonic irrespective of the value of $t \neq 0$.

ii) If $R > 0$ then for $t^2 = \frac{12}{R}$, f with parallel mean curvature implies that $\varphi_-(\varphi_+)$ is harmonic, while for $t^2 \neq 0$, $\frac{12}{R}$ the conclusion of i) holds.

Remark

The first part of ii) can be interpreted as a Ruh–Vilms type theorem on the harmonicity of the Gauss map, see [19], [20], [21].

We conclude by observing that from (3.6) we obtain the formulas

$$\begin{cases} i(K - R_{1212})\varphi \wedge \overline{\varphi} = \varphi^*_-(k_+) + \varphi^*_+(k_+) - i\varphi \wedge \overline{\varphi} = -\varphi^*_-(k_-) - \varphi^*_+(k_-) + i\varphi \wedge \overline{\varphi} \\ i(K^\perp - R_{1234})\varphi \wedge \overline{\varphi} = \varphi^*_-(k_+) - \varphi^*_-(k_-) = \varphi^*_+(k_-) - \varphi^*_-(k_-) \end{cases} \tag{3.11}$$

for $K, K^{\perp}$ the Gaussian and normal curvatures of $f : M \to N$. (3.11) are the exact analogue of (1.4) in this more general situation. For instance, if we take $N = S^4$ with its canonical metric then it is known that (W, ds_t^2, J_+) behaves essentially like CP^3, an application of the Chern–Gauss–Bonnet theorem to (3.11) for M compact gives

$$\begin{aligned} \chi(M) &= c\{\deg(\varphi_-) + \deg(\varphi_+)\} \\ \chi(TM^{\perp}) &= c\{\deg(\varphi_-) - \deg(\varphi_+)\} \end{aligned} \tag{3.12}$$

where the normalizing non–zero constant c depends on the dilatation factor t. Of course (3.12) are the analogues of (1.5), (1.6) and as a consequence

Theorem 3.2

Let $f : M \to S^4$ be an isometrically immersed compact Riemann surface with twistor lifts $\varphi_{\pm} : M \to (W, ds_t^2, J_+)$. Then M is homeomorphic to a torus if and only if $\deg(\varphi_-) = -\deg(\varphi_+)$ *and the normal bundle has zero Euler characteristic if and only if* $\deg(\varphi_-) = \deg(\varphi_+)$.

The only missing point in the entire picture seems to be the Eisenhart representation (1.3). Of course we do not expect to be able to write such an integral formula, now its role is played by the twistor projection T, roughly speaking given $\varphi : M \to W, J_-$ holomorphic, one can descend to N via the composition $T \circ \varphi$. This has been done by Eells and Salamon [13] and we refer to their paper for the pertinent statement of the result (in particular Corollary 5.4).

REFERENCES

[1] M. Gell–Mann and M. Levy, Nuovo Cimento **16** (1960) 53.

[2] J. Eells and L. Lemaire, *A report on harmonic maps*, Bull. L.M.S. **10** (1978) 1–68.

[3] A.M. Din and W.J. Zakrewski, *General classical solutions in the* $\mathbf{C}P^{n-1}$ *model*, Nucl.Phys. **B174** (1980) 397–406.

[4] J. Eells and J.C. Wood, *Harmonic maps from surfaces to complex projective spaces*, Advances in Math. **49** (1983) 217–263.

[5] D. Burns, *Harmonic maps from* $\mathbf{C}P^1$ *to* $\mathbf{C}P^n$, Proc. Tulane Conf. Lecture Notes in Math. **949** (Springer–Verlag, 1982) 48–56.

[6] F.M. Atiyah, *Instantons in two and four dimensions*, Comm.Math.Phys. **93** (1984) 437–451.

[7] S.S. Chern, *Minimal surfaces in Euclidean space of N dimensions*, Symposium in honour of Marston Morse (Princeton University Press, 1965) 187–198.

[8] W. Blaschke, *Sulla geometria differenziale delle superfici* S_2 *nello spazio euclideo* S_4 Ann.Mat.Pura Appl.(4) **28** (1949) 205–209.

[9] D.A. Hoffman and R. Osserman, *The Gauss map of surfaces in* $\mathbf{R}^3$ *and* $\mathbf{R}^4$, Proc. L.M.S. **50** (1985) 27–56.

[10] D.A. Hoffman and R. Osserman, *The geometry of generalized Gauss map*, Memoires A.M.S. **236** (1980).

[11] G.R. Jensen and M. Rigoli, *Surfaces in 4 manifolds*, in preparation.

[12] F.M. Atiyah, N.J. Hitchin and I.M. Singer, *Self-duality in four dimensional Riemannian geometry*, Proc.Roy. Soc. London Ser.A **362** (1978) 425–461.

[13] J. Eells and S. Salamon, *Twistorial constructions of harmonic maps of surfaces into four-manifolds*, Ann. Scuola Normale Superiore di Pisa (4) **12** (1985) 589-640.

[14] I.M. Singer and J.A. Thorpe, *The curvature of 4-dimensional Einstein spaces*, in Global Analysis in honour of Kodaira (Princeton University Press, 1969) 355–365.

[15] S. Salamon, *Topics in four dimensional Riemannian geometry* in Geometry Seminar "Luigi Bianchi", Ed. E. Vesentini, Lecture Notes in Math **1022** (Springer Verlag, 1983) 33–124.

[16] T. Friedrich and H. Kurke, *Compact four-dimensional self dual Einstein manifolds with positive scalar curvature*, Math. Nachr. **106** (1982) 271–299.

[17] W.P. Thurston, *Some simple examples of symplectic manifolds*.Proc. A.M.S. **55** (1976) 467–468.

[18] A. Lichnerwicz, *Applications harmoniques et variétés Kählleriennes*, Symp. Math. III Bologna (1970) 341–402.

[19] E.A. Ruh and J. Vilms, *The tension field of the Gauss map*, Trans. A.M.S. **149** (1970) 569–573.

[20] C.M. Wood, *The Gauss section of Riemannian immersion*, Journal L.M.S. **33** (1986) 157–168.

[21] G.R. Jensen and M. Rigoli, *Harmonic Gauss maps*, in preparation.

II: Théories de jauge et des monopoles

FIBRATION DE HOPF ET LA TOPOLOGIE DES MONOPOLES

P. A. HORVÁTHY

Département de Mathématiques, Université de Metz,
F-57045 METZ, France.

Résumé: L'argument de la quantification de la charge électrique en présence d'un monopole de Dirac est revu et relié à la fibration de Hopf. Les théories Grand-Unifiées (formulées aussi en termes de fibrés) admettent des solutions (étudiées pour la première fois par 't Hooft et Polyakov), qui, toute en étant partout régulières et ayant une énergie finie, ressemblent au champ singulié postulé par Dirac. Dans ces théories les charges électriques et magnétiques satisfont à des conditions de quantification généralisées, permettant d'inclure aussi des particules à charge fractionnelle.

Abstract: The argument leading to electric charge quantization in a Dirac monopole field is reviewed and linked to the Hopf fibration. The Grand Unified Theories (formulated also in fibre bundle terms) admit solutions (first studied by 't Hooft and Polyakov) which, while being everywhere regular and having finite energy, are similar to Dirac's singular field for large distances. In these theories the electric and magnetic charges satisfy generalized quantization conditions, accomodating also fractionally charged particles.

1. INTRODUCTION

Une des grands énigmes de la physique est la quantification de la charge électrique: toute charge observée est un multiple entier de celle de l' électron. C'est justement pour expliquer ce phenomène fondamental que Dirac [1] a postulé, en 1931, l' existence d'une particule hypothétique qui serait la source d'un champ magnétique radial. Il a montré en effet qu' une particule chargée dans un tel champ admet une description quantique admissible si et seulement si g, la charge du monopole et q, la charge électrique de la particule d'essai, satisfont à la relation

$$2qg = n, \quad \text{un entier} \tag{1.1}$$

D'autre part, c'est cette même année que Hopf [2] a introduit la fibration

$$U(1) \rightarrow S^3 \rightarrow S^2 \tag{1.2}$$

qui porte aujourd'hui son nom et qui est le premièr fibré non-trivial étudié en mathématiques.

La relation de ces deux constructions est tout à fait remarquable: comme on l'a noté d'abord en quantification géometrique [3,4] et accepté finalement par des physiciens [5], l'électromagnétisme peut être décrit en terme d'une forme de connection A sur un fibré en cercle au dessus de l'espace - temps. (Voir Références 6, 7 pour la terminologie). L'argument [8] (motivé par l' approche intuitive de Feynman [9]) est revu dans notre premier chapitre.

Le monopole de Dirac possède cependant un certain nombre de défauts: il a été introduit d'une manière *ad hoc*, il a une énergie infinie, et, surtout, il n'a jamais été observé par les expérimentateurs.

Début des années 1970 sa *raison d'être* semblait être menacée: c'était l'avènement des théories Grand-Unifiées, qui ont fourni une explication naturelle à la quantification de la charge électrique, sans devoir faire appel à un objet aussi singulier que le monopole.

Or, c'est justement dans ces théories Grand-Unifiées que 't Hooft, et Polyakov [10] ont montré l'existence de solutions exactes aux équations de champ, qui, à longues distances, ressemblent au champ du monopole de Dirac tout en étant partout réguliers et ayant une énergie finie. Si nous croyons à la pertinence physique des théories Grand-Unifiées nous sommes amenés à croire à l'existence des solutions exactes donc des monopoles. (Les preuves expérimentales font, hélas, toujours défaut).

Au chapitre 3. nous étudions brièvement les configurations Yang-Mills Higgs statiques et présentons la solution t' Hooft- Polyakov. (Pour plus de détails le lecteur est invité à consulter les revues [11,12]).

Le lien avec le monopole de Dirac est établi par le théorème de Goddard, Nuyts et Olive [13] : à grande distances, tout monopole est un monopole de Dirac plongé dans la théorie Grand-Unifiée. Ceci s'exprime en termes de réductions de fibrés.

Nos monopoles Grand-Unifiés possèdent des propriétés topologiques remarquables: elles tombent dans des secteurs séparés de barrières infinies d'énergie, qui correspondent à des classes d'homotopie (Cette même classification est obtenue, une fois de plus, en termes de fibrés [6,7]).

Une des questions qu'on se pose en théorie Grand-Unifiée est la suivante: dans l' équation (1.1) : qu' est la charge électrique? qu' est la charge magnétique ? (et par conséquant, ce que c'est "n" ?) . En d'autres mots: comment la condition de Dirac (1.1) doit être modifiées dans une théorie Grand-Unifiée? Une telle modification est nécessaire, si on veut que la théorie contienne des particules ayant une charge électrique fractionnelle, dont l exemple serait un "quark".

Les charges magnétiques et électriques sont introduites au chapitre 5. Quelques hypothèses supplémentaires permettent alors de démontrer une nouvelle condition de quantification

$$2q_{min}\, g_{min} = 1/M, \quad M \text{ étant un entier.} \tag{1.3}$$

Dans certains cas nous pouvons avoir $M > 1$, ce qui permet de concilier l'existence de charges fractionnelles et des monopoles magnétiques [14,15].

2. LE MONOPOLE DE DIRAC

Il est pratique de formuler l'électromagnétisme en utilisant le langage des formes différentielles: le champ électromagnétique (E, B) est décrit par une 2-forme **F** définie sur l'espace-temps,

$$\mathbf{F} = (1/2)\, F_{\mu\nu}\, dx^{\mu} \wedge dx^{\nu} \quad \text{où} \quad F_{i0} = E_i\,, \quad F_{ij} = \varepsilon_{ijk} B_k \tag{2.1}$$

en termes de laquelle les équations de Maxwell s'expriment comme

$$d\,\mathbf{F} = 0 \quad \text{et} \quad d{*}\mathbf{F} = 0 \tag{2.2}$$

où $*\mathbf{F}$ est le dual de Hodge de la 2-forme **F**. Un potentiel pour **F** est une 1-forme **A** telle que d **A** = **F**. **A** est relié aux potentiels scalaires V et A par $\mathbf{A} = - Vdt + A_i\, dx^i$.

En 1931 Dirac [1] postula l'existence d'un champ magnétique radial

$$B = g\mathbf{r}/r^3. \tag{2.3}$$

La 2-forme correspondante est proportionnelle à l'image réciproque par la projection p: QxR → S^2, $p(\mathbf{r},t) \to \mathbf{r}/r$ de la forme de surface canonique Ω de la deux-sphère,

$$\mathbf{F} = g\, p^{*} \Omega = (g/r^3)\, \mathbf{r}.d\mathbf{r}{\times}d\mathbf{r} \tag{2.4}$$

Ce champ vérifie les équations de Maxwell dans le vide partout, à l'exception de l'origine. En particulier, **F** est <u>fermée</u>, d **F** = 0. Cependant, ce champs <u>ne dérive pas</u> d'un potentiel-vecteur: l'hypothèse **F** = d **A** (c'est à dire B = rot A) mènerait en effet à une contradiction: si S^2 est une deux-sphère arbitraire de l'espace qui contient l'origine à son intérieur, le théorème de Stokes nous donnerait

$$\int_{S^2} \mathbf{F} = \int_{S^2} d\,\mathbf{A} = \int_{\partial S^2} \mathbf{A} = 0, \tag{2.5}$$

car S^2 n'a pas de bord. Or le calcul direct nous donne

$$\int \mathbf{F} = 4\pi g, \tag{2.6}$$

une contradiction (L'équation (2.6) nous fournit aussi l'interprétation de g comme la charge magnétique, $g = (1/4\pi) \int \mathbf{F}$.)

Quelles sont les <u>conséquences physiques</u> de ce fait ? Au niveau de la mécanique classique aucune, puisque les équations de mouvement d'une particule d'essai sont exprimées en termes de **B**:

$$m\ddot{\mathbf{r}} = q\,\mathbf{B} \times \mathbf{v}. \tag{2.7}$$

Il y a, par contre, des conséquences au niveau <u>quantique</u>. L'objet fondamental à regarder est en effet le <u>propagateur</u> $K(x,t|\,x',t')$, exprimé comme un "intégral sur les chemins", symboliquement

$$K(x,t\,|\,x',t') = \int_{\wp} \exp\,[(i/\hbar)\,S(\gamma)]\,D\gamma \tag{2.8}$$

où $\wp$ denote tous les chemins dans l'espace-temps qui commencent à $x' = (\mathbf{r}',t')$, finissent à $x = (\mathbf{r},t)$. $S(\gamma) = \int_{\gamma} L$ est <u>l'action classique</u> calculée le long du chemin γ, L étant le <u>Lagrangien</u> du système. Pour une particule soumise à un champ électromagnétique ce dernier est donné par

$$L = mv^2/2 + q\,(\mathbf{A}.\mathbf{v} - V) \tag{2.9}$$

Pour une particule dans le champ d'un monopole de Dirac la définition même de l'action classique est problématique, faute d'un potentiel-vecteur défini partout. La condition $d\,\mathbf{F} = 0$ implique cependant, par le lemme de Poincaré, qu'un potentiel-vecteur existe dans tout ouvert contractile, comme dans $U_+ = \mathbf{R}^3 \setminus \{$le demi-axe des $z > 0\}$. En coordonnées polaires nous avons en effet

$$A_r = A_\theta = 0\,,\ A_\phi^+ = g(1-\cos\theta). \tag{2.10}$$

Pour des chemins γ qui ne rencontrent pas la "corde" $z > 0$ nous pouvons définir alors l'action classique en intégrant L_+ avec $\mathbf{A}_+$. Pour des chemins qui coupent la corde nous pouvons choisir un autre ouvert, par exemple $U_- = \mathbf{R}^3 \setminus \{$le demi-axe des $z < 0\}$, où nous avons le potentiel-vecteur

$$A_r = A_\theta = 0\ ,\ A_\phi^- = g(-1-\cos\theta) \tag{2.11}$$

Dans l'intersection des deux ouverts nous disposons alors de deux descriptions différentes et en particulier de deux formules différentes pour l'action classique. Quand devons - nous les considérer comme équivalents? Observons que l'action classique est toujours ambigüe : on peut toujours rajouter une dérivée totale au Lagrangien, $L \to L + df/dt$, qui change l'action par la constante $f(x) - f(x')$. Son effet est de multiplier le "facteur de Feynman"

$$\exp[(i/\hbar)S] \to c(x,x') \exp[(i/\hbar)S]\ , \tag{2.12}$$

(et par conséquant l'évolution quantique) par le facteur de phase $c(x,x') = \exp i[f(x)-f(x')]$ qui ne dépend que des points d'extrémité et non pas du chemin γ particulier et qui est ainsi inobservable. En adoptant la philosophie que c'est justement l'ambiguité maximale permise, nous allons demander qu'en remplaçant un potentiel-vecteur par un autre le facteur de Feynman ne change que par un facteur de phase, comme en (2.12).

Soit maintenant $A^{(+)}$ et $A^{(-)}$ deux potentiels définis dans des ouverts U_+ et U_- et γ_1 et γ_2 deux chemins entre les points x et x' de l'intersection. L'équivalence physique des deux descriptions exige que les facteurs de Feynman calculés à l'aide de $A^{(-)}$ et $A^{(+)}$ soient reliées comme dans (2.12); en divisant les deux expressions (2.12) pour γ_1 et γ_2 le facteur $c(x,x')$ disparait et nous voyons que ceci revient, compte tenu de la forme (2.9) de l'action, à demander

$$\exp[(iq/\hbar)\oint A^{(+)}] = \exp[(iq/\hbar)\oint A^{(-)}] \tag{2.13}$$

où les integrations sont le long du lacet $\gamma = (\gamma_1)^{-1}\gamma_2$. Ceci s'écrit aussi comme

$$\exp[(iq/\hbar)\oint(A^{(-)} - A^{(+)})]. \tag{2.14}$$

En "mettant des chapeaux" dans U_+ et dans U_- ayant le lacet γ pour bord, appliquant le théorème de Stokes dans les deux cas, tenant compte des orientations, la condition (2.14) s'exprime aussi comme

$$\int \mathbf{F} = (q/2\pi\hbar)\, n, \quad n \text{ un entier}, \tag{2.15}$$

soit, compte tenu de (2.6),

$$2qg/\hbar = n, \quad \text{un entier.} \tag{2.16}$$

(2.16) est la fameuse condition de quantification de Dirac: s'il existe un monopole de charge g dans l' univers, toute charge électrique doit être un multiple entier de

$$q_{min} = \hbar/2g \tag{2.17}$$

Le lemme de Weil [3] permet de reformuler ces résultats en termes de fibrés. La condition (2.15) est en effet la condition nécessaire et suffisantes pour qu'il existe un <u>fibré principal</u> Y au dessus de l'espace-temps $(S^2 \times R^+) \times R$, ayant le cercle U(1) pour groupe structural, muni d'une <u>forme de connection</u> $\mathcal{A}$ dont la courbure est $q\mathbf{F}/\hbar$. Explicitement, soit

$$Y_\pm = U_\pm \times U(1) = \{(x, z_\pm)\}. \tag{2.18}$$

La condition (2.14) nous garanti que

$$g(x) = \exp\left[(iq/\hbar)\int_{r_0}^{r} (A^{(+)} - A^{(-)})\right] \tag{2.19}$$

(où l'intégration est le long d'un chemin γ arbitraire qui relie le point de référence $\mathbf{r}_0$ à $\mathbf{r}$) est indépendant du choix de γ et défini ainsi une fonction sur $U_+ \cap U_-$. En choisissant $\mathbf{r}_0$ égale au "pôle est" sur la sphère unité, $\mathbf{r}_0$ = (1,0,0) en coordonnées polaires (r, θ, ϕ), la forme explicite (2.10) nous montre qu'en effet

$$g(x) = e^{i(2qg/\hbar)\phi} = e^{in\phi}. \tag{2.20}$$

Appelons maintenant deux points (x_+, z_+) et (x_-, z_-) dans $Y_\pm$ équivalents si et seulement si

$$x_+ = x_- = x \quad \text{et} \quad z_+ = g(x) z_- \tag{2.21}$$

Soit Y le quotient de $\{Y_+, Y_-\}$ par cette relation d'équivalence. Le cercle U(1) opère sur Y selon $[(x,z)] \to [(x, zc)]$, $c \in U(1)$. $y = [x,z] \to \mathbf{x}$ défini finalement une projection de Y sur l'espace-temps.

Les potentiels $A^{(+)}$ et $A^{(-)}$ sont reliés commes

$$A^{(+)} = A^{(-)} + (\hbar/iq)\, dg/g. \qquad (2.22)$$

Définissons maintenant les 1-formes

$$\mathcal{A}^{(\pm)} = A^{(\pm)} + (\hbar/iq)\, dz_\pm/z_\pm. \qquad (2.23)$$

sur $Y_\pm$. (2.22) montre que ces définitions sont compatibles et définissent ainsi une 1-forme *globale* $\mathcal{A}$ sur Y. Elle est de plus invariante par l'action de U(1) et prend la valeur 1 sur le générateur infinitesimal $(0, iq/\hbar\, z)$ de cette action. Ce sont exactement les propriétés qu'on demande à une forme de connexion.

Une fonction continue s définie sur un ouvert U de l'espace-temps telle que $\pi(s(x)) = x$ est appelé une section du fibré Y. En termes physiques choisir une section revient à choisir un <u>jauge</u>.

L' image réciproque d'une forme de connexion $\mathcal{A}$ par une section locale,

$$A^s = s^* \mathcal{A} \quad , \qquad (2.24)$$

est un potentiel local pour $q\mathbf{F}/\hbar$, $q\mathbf{F}/\hbar = dA^s$, car dg/ig est un 1-forme fermée.

Rappelons - nous maintenant de la définition du transport parallèle: un vecteur tangent $\bar{X}$ à Y au point y est horizontal si $\mathcal{A}(\bar{X}) = 0$. Localement, $\bar{X} = (X, -(iq/\hbar)A(X)z)$. Un chemin $\bar{\gamma}$ est appelé horizontal si son vecteur tangent l'est dans chaque point. Si γ est un chemin dans l'espace-temps, et y un point arbitraire dans la fibre au desus de $\gamma(o)$, il existe un chemin horizontal unique $\bar{\gamma}$ qui passe par y, appelé le relèvement horizontal de γ. En coordonnés locales $\bar{\gamma}(t) = (\gamma(t), z(t))$ où

$$z(t) = z(o) \exp -[(iq/\hbar) \int_\gamma A] \qquad (2.25)$$

Ceci permet de définir l'action classique pour tout chemin dont les extrémités sont contenues dans le domain d'un potentiel; une telle définition satisfait à la condition (2.12), voir [8].

Nous pouvons considérer en particulier les lacets γ_ϕ, $0 \leq \phi \leq 2\pi$ obtenues en suivant le méridien $\phi = 0$ du pôle du nord N jusqu'au pôle du sud S, et en remontant à N le lond du méridien à l'angle ϕ ;. Observons que

$$h(\phi) = \exp -[(iq/\hbar) \oint_\gamma A] \tag{2.26}$$

est un <u>lacet</u> dans U(1), qui ne dépend pas du choix du potentiel A. Sa <u>classe d' homotopie</u> [h] est exactement l' entier n définie par la condition de Dirac (2.16). L'expression (2.10) donne en effet

$$h(\phi) = e^{in\phi}. \tag{2.27}$$

Ceci permet d'interpreter le nombre "quantique" n comme le nombre de tours faite par $h(\phi)$, quand ϕ passe de 0 à 2π.

Ce que nous venons de démontrer est en effet la classification des fibrés principaux au dessus de la sphère [7].

Dans le cas du monopole de Dirac, nous disposons de formules explicites. En "oubliant" les variables r,t nous allons travailler sur la sphère. Considérons la sphère unité S^3 de l'espace C^2:

$$S^3 = \{\zeta = (z^1, z^2) \in C^2 \mid |z_1|^2 + |z_2|^2 = 1\}. \tag{2.28}$$

U(1) opère sur S^3 selon $(z_1, z_2) \to (z_1 c, z_2 c)$, $c \in U(1)$. Nous avons en particulier l' action du sous-groupe discret $Z_n = \{ \exp 2\pi k/n, 0 \leq k \leq n-1 \}$; soit

$$Y_n = S^3 / Z_n . \tag{2.29}$$

Y_n est un fibré principal ayant pour base S^2, la projection étant donnée par

$$(\pi([\zeta]))_i = \bar{\zeta} \sigma_i \zeta, \qquad i = 1,2,3, \tag{2.30}$$

où les σ_i sont les matrices de Pauli.

$$\mathcal{A}_n = n \bar{\zeta} d\zeta / i = (n/i)(\bar{z}_1 dz_1 + \bar{z}_2 dz_2) \tag{2.31}$$

est une forme de connexion sur Y_n dont la courbure est n Ω, la forme de surface de la sphère, multipliée par l'entier n.

En choisissant des angles d' Euler, un point de S^3 = SU(2) s'écrit comme

$$\zeta = \begin{cases} \exp[i(\phi+\chi)/2] \cos\theta/2 \\ i \exp[-i(\phi-\chi)/2] \sin\theta/2 \end{cases} \tag{2.32}$$

La classe d'équivalence de

$$s_+(\theta,\phi) = \begin{cases} \cos\theta/2 \\ i\exp[-i\phi] \cos\theta/2 \end{cases} \tag{2.33}$$

est alors une section locale du fibré Y_n qui est définie sur $U_+ = \{\theta,\phi \mid 0 \leq \theta < \pi\}$. La substitution de s_+ dans (2.31) montre que l'image réciproque de la forme de connection $\mathcal{A}_n$ est exactement le potentiel (2.10).

Nos résultats se résument de la manière suivante: le champ électromagnétique est représenté par une forme de connexion sur un fibré principal en U(1) au dessus de l' espace-temps. Dans le cas du monopole de Dirac ce fibré est caractérisé par un entier n, déterminé par la quantification des charges électriques et magnétiques. Le fibré en question n'est trivial que pour n = 0.

3. MONOPOLES GRAND-UNIFIES

Considérons d'abord les ingrédients d'une théorie Grand-Unifiée générale. Soit G (appelé le <u>groupe de jauge</u>) un groupe de Lie compact et connexe, et P un fibré principal au dessus de $\mathbf{R}^3$ ayant G pour groupe structural. Un tel fibré est nécessairement trivial, car $\mathbf{R}^3$ est contractile. Il s'identifie alors au produit

$$P = \mathbf{R}^3 \times G, \tag{3.1}$$

l'action de G étant définie par $p = (x,g) \rightarrow (x, gh) = ph$, $h \in G$. Une <u>transformation de jauge</u> est un automorphisme du fibré qui commute avec l'action (à droite) de G sur P, et qui se projette sur l'identité de la base. Compte tenu de la forme (3.1), elle se met sous la forme

$$p = (x,k) \rightarrow (x, g(x)k). \tag{3.2}$$

Une section de P (une "choix de jauge" des physiciens) est une application s de la base dans le fibré tel que s(x) est dans la fibre au dessus de x.

Un <u>champ</u> de <u>Yang-Mills</u> (statique) est une connexion sur P, représentée par une 1-forme $\mathcal{A}$ à valeurs dans l'algèbre de Lie $\mathcal{G}$ de G, équivariante par rapport à l'action de G et qui est proprement normalisée par l' action infinitésimale. (voir, par exemple, Kobayashi et Nomizu [6]).

L'image réciproque d'un champ de Yang-Mills par une section s est une 1-forme $A^s = s^*\mathcal{A} = A_i dx^i$ qui satisfait à loi de transformation

$$A^s \rightarrow \mathrm{Ad}\, g^{-1} A^s + g^{-1}dg \tag{3.3}$$

par rapport à une transformation de jauge (3.2).

Le deuxième ingrédient est un champ scalaire appelé <u>champ de Higgs</u> Soit en effet V un espace de représentation (de dimension fini) pour G muni d'un produit scalaire invariant par G. Un champ de Higgs est une fonction equivariante sur P à valeurs dans V,

$$\Phi : P \rightarrow V, \quad \Phi(pg) = g^{-1}.\Phi(p) \tag{3.4}$$

où g.v denote l'action de G sur V. En terme locaux $\Phi^s(x) = \Phi(s(x))$ est une fonction qui a la loi de transformation

$$\Phi^s = g^{-1}.\Phi^s \tag{3.5}$$

Par exemple, on peut prendre $V = \mathcal{G}$. G opère selon $\xi \to \mathrm{Ad}\, g\xi$ (représentation adjointe); un produit invariant est donné par $(\xi,\eta) = \mathrm{Tr}(\xi\eta)$.

Soit finalement U (appelé un <u>potentiel de Higgs</u>) une fonction non négative sur V qui est invariante par G:

$$U \geq 0, \quad U(gv) = U(v)\,, \tag{3.6}$$

Les minima (supposés 0) de U sont alors invariants par G. Nous faisons l'hypothèse supplémentaire, que G opère transitivement sur l'ensemble des minima de U qui devient de cette manière une orbite de G dans V, qu'on peut identifier à G/H, H étant un sous-groupe compact de G, appelé le <u>groupe résiduel</u>.

L'<u>énergie</u> d'une configuration statique ($\mathcal{A},\Phi$) est donnée par

$$E = \int \{ \frac{1}{2} |F|^2 + \frac{1}{2}(D\Phi, D\Phi) + U(\Phi) \}\, d^3x \tag{3.7}$$

où $|F|^2 = \mathrm{tr}\, F \wedge * F = (1/2) F_{ij}F^{ij}$, $F = (1/2)F_{ij}\, dx^i \wedge dx^j$ étant la courbure de la connexion $\mathcal{A}$

$$F_{ij} = \partial_i A_j - \partial_j A_i + [A_i, A_j]. \tag{3.8}$$

Physiquement, F est le tenseur du champ de jauge.

Un monopole Grand-Unifié est une solution exacte à énergie finie aux équations variationnelles associées à (3.7), qui s'expriment comme

$$*D* F = - \{\Phi, D\Phi\}, \qquad *D *D\Phi = - \delta U/\delta\Phi \tag{3.9}$$

ou localement comme

$$(D_i F^{ij})_a = (\Phi, \tau_a D_j \Phi), \qquad D_i D^i \Phi = - \delta U/\delta\Phi, \quad a = 1,...\dim G, \; i,j = 1,2,3. \tag{3.9'}$$

ce qui definit aussi l'opération $\{.,.\}$: $V \times V \to \mathcal{G}$. Pour $V = \mathcal{G}$, $\{\Phi, D\Phi\} = [\Phi, D\Phi]$.

Le cas étudié par 't Hooft et Polyakov [10] correspond à

$$G = SU(2), \; V = su(2) \approx \mathbf{R}^3, \; U(\Phi) = \lambda(\mu^2 - \Phi^2)^2 \Rightarrow \Phi_0 = \mu\sigma_3, \; H = U(1)). \tag{3.10}$$

't Hooft et Polyakov ont montré en effet que l'Ansatz

$$A_i^a = \varepsilon_{iaj} x^j/r^2 \,(K(r) - 1), \qquad \Phi^a(r) = (x^a/r)H(r) \tag{3.11}$$

contient une solution à énergie finie.

Les seules solutions explicites connues à ce jour [16,17] correspondent à $\lambda = 0$ (limite de "Prasad - Sommerfield"); dans le cas (3.11) c'est

$$H(r) = \mu r\,(1 - \coth \mu r), \quad K(r) = \mu r / \mathrm{sh}\, \mu r \tag{3.12}$$

La masse (l' énergie) de cette configuration est

$$E = 4\pi\, \mu = (1/\alpha)\, M_W \tag{3.13}$$

où M_W est la masse du vecteur-boson, et α est la constante de la structure fine et dont la valeur est approximativement 1/137 [11,12].

La masse de l'objet de 't Hooft- Polyakov ($\lambda \neq 0$) est encore légèrement supérieure, ce qui explique pourquoi les tentatives de produire des monopoles dans des collisions à haute énergie sont restées sans succès: Il fallut en effet un effort (et un budget!) incroyable aux observateurs du CERN pour mettre en évidence les W et Z du modèle Weinberg-Salam, et dont les masses M_W ne sont "que" 100 fois supérieurs à celle du proton.

4. ASPECTS TOPOLOGIQUES

Les monopoles Grand-Unifiés admettent des propriétés topologiques remarquables. Pour le voir, considérons les conditions de convergence de l'intégrale (3.7) dans le cas général.

A.) Le fibré P étant trivial, le champ de Higgs Φ s'identifie, d'une manière canonique, avec une fonction sur $\mathbf{R}^3$, que nous noterons encore Φ. Pour $\int U(\Phi)d^3x < \infty$ il est nécessaire qu' à grandes distances le champ de Higgs prenne ses valeurs dans l'orbite **G/H** des minima du potentiel de Higgs U. Φ défini par conséquent une application d'une grande sphère S^2 - appelée la "sphère à l'infini" - dans l'orbite:

$$\Phi : S^2 \rightarrow G/H \tag{4.1}$$

B.) $\int (D\Phi)^2 \, d^3x < \infty$ est assuré par

$$|D\Phi| = O(r^{-2}) \quad \Rightarrow \quad D\Phi = 0 \quad \text{sur } S^2 \tag{4.2}$$

C.) Pour $\int |\mathbf{F}|^2 d^3x < \infty$ nous demandons

$$\mathbf{F} = F/r^2 , \tag{4.3}$$

F étant la courbure d'une connexion sur S^2.

En termes de fibrés ces propriétés s'expriment comme

A'). $P|S^2$, la restriction du fibré (3.1) à la sphère à l'infini, doit se réduire à un fibré principal Q ayant le groupe résiduel H pour groupe structural [6,7]. Ce fibré est isomorphe à

$$Q = \Phi^{-1}(\Phi_0) \ , \quad \text{où} \quad \Phi_0 \text{ est un point de l'orbite } G/H. \tag{4.4}$$

B'). (4.2) entraine que la connexion de Yang-Mills $\mathcal{A}$ se réduit à une connexion sur (Q,H).

C'). F est la courbure de la connexion réduite. A grandes distances. Les équations de champ (3.9) se découplent; elles prennent la forme asymptotique

$$D\,\Phi = 0 \ , \quad D{*}F = 0 \quad \text{i.e.} \quad D_j\,\Phi = 0, \ D_i F^{ij} = 0 \text{ sur } S^2. \tag{4.5}$$

Analysons maintenant ces conditions plus en détails.

A.) La relation (4.1) nous définit une <u>classe d'homotopie</u>

$$[\Phi] \in \pi_2(G/H) \tag{4.6}$$

qui est invariante par les déformations continues des champs [12]. On montre aussi, que Φ et Φ' sont reliés par une transformation de jauge

$$\Phi(x) = g^{-1}.\Phi(x) \tag{4.7}$$

si et seulement si $[\Phi] = [\Phi']$. En d' autres termes, ssi les réductions (Q,H) et (Q',H) définies par (4.4) sont équivalentes. Le fibré (Q,H), ayant la variété non-contractile S^2 pour base, peut ne pas être trivial; il est trivial ssi $[\Phi] = 0$.

L'invariant (4.6) s'appelle la <u>charge topologique</u> ou la <u>charge de Higgs</u>. Le raisonnement ci-dessus montre que deux configurations ayant des charges topologiques différentes sont séparées par des barrières infinies d'énergie.

Pour $G/H = SU(2)/U(1) = S^2$, $\pi_2(S^2) = \mathbf{Z}$; la charge topologique s'identifie donc à un entier m qui compte le nombre de fois Φ couvre l'image - S^2. Dans le cas 't Hooft-Polyakov ce nombre est unité. La construction (dans la limite Prasad-Sommerfield $\lambda = 0$) de solutions exactes à charge topologique $m > 0$ [17] a été un des grands exploits de la physique mathématique des années récentes.

La charge topologique peut aussi être caractérisée par des classes en $\pi_1(H)$, le premier groupe d'homotopie du groupe résiduel. En effet, G opère sur l'orbite G/H, et nous pouvons chercher un relèvement, c'-est-à-dire une fonction g(x) sur S^2 à valeurs dans G tel que

$$\Phi(x) = g(x).\Phi_0 \tag{4.8}$$

Un relèvement continu sur la sphère entière n'existe que si $[\Phi] = 0$; il existe cependant sur tout ouvert contractile U, par exemple sur les U_+ et U_- définis au Chapitre 2. Dans l'intersection $U_+ \cap U_-$ nous disposons alors de deux relèvements g_+ et g_- qui diffèrent, par (4.8), par une fonction à valeurs dans le groupe résiduel H. La restriction de

$$h(x) = g_+^{-1}(x) g_-(x) \tag{4.9}$$

à l'équateur S^1 est par conséquant un <u>lacet dans H</u>. Sa classe dans $\pi_1(H)$ est exactement l'image de $[\Phi]$ par l'homomorphisme injectif $\delta : \pi_2(G/H) \to \pi_1(H)$.

De plus le lacet (4.9) est contractile dans G, puisque $P|S^2$ est trivial. Ceci est consistant avec le résultat [7] qui dit que les fibrés principaux aux dessus de la sphère sont caractérisés par les classes dans π_1.

La relation $\pi_1(U(1)) = \mathbb{Z}$ montre aussi que notre second type de description est consistant avec la precedente en termes de π_2 pour 't Hooft - Polyakov.

Le π_1 de tout groupe de Lie compact peut être décrit [15]. Dans ce qui suit nous ne considérons que le cas $\pi_1 = \mathbb{Z}$, qui présente un intérêt physique particulier.

Le modèle le plus simple est celui de 't Hooft-Polyakov étudié ci-dessus. Un autre exemple simple, qui contient cependant les aspects du cas général, est donné par [18]

$$G = SU(3), \quad V = \mathcal{G} = su(3), \quad \Phi_0 = \mathrm{diag}\,(1,1,-2) \Rightarrow$$

$$H = U(2) = SU(2)xU(1)]/\mathbb{Z}_2\,, \; G/H = P_2(\mathbb{C}); \quad \pi_1(H) = \pi_2(G/H) = \mathbb{Z}. \qquad (4.10)$$

On montre [15] que demander $\pi_1(H) = \mathbb{Z}$ revient à demander que z, le centre de l'algèbre de Lie $\mathfrak{h}$, soit uni-dimensionnel et $\mathbf{H_{ss}}$, le sous-groupe semi-simple de H engendré par $[\mathfrak{h}, \mathfrak{h}]$, soit simplement connexe. Soit $\mathbf{z}$ la projection de $\mathfrak{h}$ sur son centre z. Les vecteurs ξ de $\mathfrak{h}$ qui vérifient $\exp 2\pi\xi = 1$ forment un réseau, appelé le réseau unité Γ. $\mathbf{z}(\Gamma)$, la projections sur z du réseau Γ est encore un réseau (unidimensionnel par l'hypothèse sur dim $\mathfrak{h}$); soit ζ son générateur. Soit de plus $\exp 2\pi\eta t$, $0 \leq t \leq 1$ un lacet dans le centre $Z(H)$ de $\mathbf{H}$. η est alors un multiple entier de ζ,

$$\eta = M\zeta, \quad M \text{ un entier}. \qquad (4.11)$$

(M est aussi le nombre de points d'intersection entre le U(1) central et le sous-groupe semi-simple $\mathbf{H_{ss}}$).

Pour 't Hooft-Polyakov H = U(1) $\Rightarrow \eta = \zeta \Rightarrow M = 1$; dans le cas (4.10)

$$\eta = \begin{pmatrix} 1 & \\ & 1 \end{pmatrix} \qquad \zeta = (1/2)\begin{pmatrix} 1 & \\ & 1 \end{pmatrix} \qquad \Rightarrow M = 2. \qquad (4.12)$$

Le lacet h (4.9) est toujours homotope à un de la forme

$$\exp 4\pi\xi t, \; 0 \leq t \leq 1, \qquad (4.13)$$

où ξ est un vecteur dans $\mathfrak{h}$. Sa projection $z(\xi)$, qui est donc un invariant topologique, est un multiple entier de ζ:

$$z(\xi) = m\zeta \tag{4.14}$$

Il a été montré [15] que la correspondance $h(t) \rightarrow m$ est un isomorphisme entre π_1 et $\mathbf{Z}$, m étant justement le nombre "quantique" caractérisant la classe.

La condition **B.**) implique que, dans un jauge convenable, le champ de Yang-Mills prend ses valeurs dans $\mathfrak{h}$. La solution de l' équation $D\Phi = 0$ est donnée par le " facteur de phase non-intégrable",

$$\Phi(x) = \mathrm{Ad}\, g(x)\, \Phi_0 \quad \text{où} \quad g(x) = \mathcal{P}(\exp - \textstyle\int \mathcal{A}), \tag{4.15}$$

où cette notation symbolique signifie "transport parallèle".

Considérons maintenant le recouvrement de $\mathbf{S}^2$ par des lacets γ_ϕ, $0 \leq \phi \leq 1$ introduit au Chapitre 2;

$$h(\phi) = \mathcal{P}(\exp - \textstyle\oint \mathcal{A}), \tag{4.16}$$

(l'intégration étant prise le long du lacet γ_ϕ) est alors un lacet dans le groupe résiduel H dont la classe d'homotopie $[h] \in \pi_1(H)$ représente le secteur topologique du monopole.

Passons finalement à l' étude de la condition **C**. La solution générale des équations de champ asymptotique (4.5) a été trouvée par Goddard, Nuyts et Olive [13]. Leur théorème dit qu' on a

$$\Phi = \Phi_0 \, , \quad A = A_D\, \rho, \tag{4.17}$$

où A_D est le potentiel de Dirac $(\pm 1 - \cos\theta)\, d\phi$, et le <u>vecteur-charge</u> ρ est un vecteur constant dans l'algèbre de Lie $\mathfrak{h}$, qui doit être quantifié,

$$\exp 4\pi\rho = 1. \tag{4.18}$$

Du point de vue physique ce théorème dit qu' asymptotiquement tout monopole est un monopole de Dirac plongé dans la direction du vecteur - charge ρ.

En termes mathématiques le théorème de GNO dit que les monopoles asymptotiques ont une holonomie U(1). Or le théorème sur la réduction d' holonomie [6] dit que tout fibré doit se réduire à son fibré d'holonomie. Le théorème de GNO garantit alors l'existence d'une plus petite réduction pour tout monopole.

D'après la formule (4.16), 2ρ appartient au réseau unité Γ, qui s' identifie au <u>réseau des charges</u>. La connaissance de ρ permet de calculer le lacet (4.16)

$$h(\phi) = e^{i2\rho\phi}, \tag{4.19}$$

ce qui montre que l'invariant topologique (4.14) est

$$2z(\rho) = 2\,\mathrm{Tr}(\rho\zeta)/\mathrm{Tr}(\zeta^2) \tag{4.20}$$

(qui est ici le seul invariant par hypothèse). Par conséquent deux monopoles dont les vecteurs - charge sont ρ et ρ' sont dans le même secteur topologique ssi ρ et ρ' ont la même projection sur le centre.

5. CHARGE MAGNETIQUE

La condition **B.)** $D\Phi = 0$ permet d'introduire d'autres invariants [15], dont le plus important est la charge magnétique. Considérons en effet la 2-forme réelle

$$\mathcal{F} = (1/\,|\Phi|\, e_0)\, \mathrm{Tr}(F\Phi) \quad \text{sur } S^2 \tag{5.1}$$

Le champ de Yang-Mills vérifie les identités de Bianchi $DF = D_i F_{ij} = 0$. Par conséquant

$$d\mathcal{F} = (1/\,|\Phi| e_0)\, \{\mathrm{Tr}(D\Phi F) + \mathrm{Tr}(\Phi D F) = 0, \tag{5.2}$$

car $|\Phi| = |\Phi_0|$ est constante sur S^2. Similairement,

$$d{*}\mathcal{F} = (1/\,|\Phi| e_0)\, \{\mathrm{Tr}(D\Phi {*}F) + \mathrm{Tr}(\Phi D {*}F) = 0 \tag{5.3}$$

compte tenu de l'équation de champ asymptotique (4.5). (5.2-3) montrent que $\mathcal{F}$ défini dans (5.1) est la courbure d'un champ réel qui vérifie les équations de Maxwell dans le vide, et peut être pris ainsi pour la définition du champ électromagnétique.

Dans le cas 't Hooft- Polyakov (3.) c'est la seule composante résiduelle: à longue distance le champ s'identifie à l'électromagnétisme. Dans le modèle (4.10), il y a quatre champs à grand rayon d'action, dont celui qui correspond au U(1) engendré par Φ s'identifie au photon de l'électromagnétisme, et ceux du sous-groupe semi-simple SU(2) sont des champs de "couleur" des interactions fortes.

Le champ (5.1) porte, en général, une charge magnétique, définie par l'integral de flux

$$g = (1/4\pi) \int \mathcal{F} = (1/4\pi e_0\, |\Phi|) \int \mathrm{Tr}(F\Phi). \tag{5.4}$$

On montre en effet que g ne dépend que de la classe topologique. Plus précisément, on démontre que

$$g = m g_{min} = m\, (|\eta|/2M), \tag{5.5}$$

Un des principes fondamentaux de la physique, la loi de superselection, demande que tous les champs physiques soient des états propres de l'opérateur de la charge electrique. En théories Grand-Unifiées ceci est donné par

$$Q_{em} = e_0\Phi/|\Phi| \,, \tag{5.6}$$

les valeurs propres q étant les charges electriques. On montre que ces dernières sont quantifiées: elles sont en effet des multiples entiers (le coefficient dépendant de la représentation selon laquelle le champ en question se transforme) d'une charge fondamentale

$$q_{min} = e_0 / |\eta| \,. \tag{5.7}$$

La comparaison des expressions (5.5) de la charge magnétique et (5.7) de la charge électrique nous montre que ces deux types de charges vérifient la <u>condition de Dirac généralisée</u>

$$2q_{min}g_{min} = 1/M \tag{5.8}$$

Dans le cas 't Hooft-Polyakov M = 1 et cette condition coincide avec la condition originale de Dirac. Pour le modèle (4.10) par contre M = 2 et la théorie permet des particules ayant une charge électrique fractionnelle ("quarks").

RÉFÉRENCES

[1] P. A. M. Dirac, Proc. Roy. Soc. (London), **A133**, 60 (1931); Phys. Rev. **74**, 817 (1948)

[2] H. Hopf, Math. Ann. **104**, (1931)

[3] J.M. Souriau, *Structure des systèmes dynamiques*, Dunod, Paris (1969); en particulier Chapître V. "Prequantification" de la nouvelle edition (1975) (non publié); B. Kostant, *Quantization and Unitary representations*, Springer Lecture Notes in Math. **170**, p. 87 (1970); D.J. Simms et N.M.J. Woodhouse, *Lectures on Geometric Quantization*, Springer Lecture Notes in Physics **53**, (1976); J. Sniatycki, *Geometric Quantization and Quantum Mechanics*, Springer Verlag, N.Y. (1980)

[4] J. Sniatycki, J. Math. Phys. **15**, 619 (1974); W. Greub, et H-R. Petry, J. Math. Phys. **16**, 1347 (1975);

[5] T.T. Wu et C. N. Yang, Phys. Rev. **D14**, 437 (1976); A. Trautman, Int. J. Theor. Phys.. **16**, 561 (1977) ; A.P. Balachandran, G. Marmo, B.S. Skagerstam et A. Stern, Nucl. Phys. **B162**, 385 (1980); J. Friedman et R. Sorkin, Phys. Rev. **D20**, 25111 (1979)

[6] S. Kobayashi et K. Nomizu : *Foundations of differential geometry* Vol. I., Interscience (1963)

[7] N. Steenrod, *Topology of fibre bundles,* Princeton Univ. Press (1951)

[8] P. A. Horvathy, dans *Differential Geometric Methods in Mathematical Physics*, Proc. '79 Conf. in Aix-en-Provence, Ed. J.M. Souriau, Springer Lecture Notes in Math. **836**, 67 (1980)

[9] R.P. Feynman et R. Hibbs, *Quantum Mechanics and Path Integrals*, McGraw-Hill, N.Y. (1965); L. S. Schulman, *Techniques and Applications of path integration*, Wiley, N.Y. (1981)

[10] G. 't Hooft, Nucl. Phys. **B79**, 276 (1974); A. M. Polyakov, JETP Lett. **20**, 194 (1974)

[11] S. Coleman " Classical lumps and their quantum descedents", Cours d'Erice, 1975; paru dans *New phenomena in subnuclear physics*, Ed. A. Zichichi, Plenum, N.Y. (1977);
"The magnetic monopole fifty years later", Cours d'Erice 1981, paru dans *The Unity of fundamental interactions*, Ed. A. Zichichi, Plenum, N. Y. (1983).

[12] P. Goddard et D. Olive, Rep. Progr. Phys. **41**, 1357 (1978)

[13] P. Goddard, J. Nuyts et D. Olive, Nucl. Phys. **B125**, 1 (1977)

[14] E. Corrigan et D. Olive, Nucl. Phys. **B110**, 237 (1976);
P. Goddard et D. Olive, Nucl. Phys. **B191**, 511 (1981)

[15] P. A. Horvathy et J. H. Rawnsley, Commun. Math. Phys. **96**, 497 (1984); **99**, 517 (1985)

[16] M.K. Prasad et C.M. Sommerfield, Phys. Rev. Lett. **35**, 760 (1975)

[17] P. Forgacs, Z. Horvath et L. Palla, Phys. Lett. **99B**, 232 (1981);
R. Ward, Commun. Math. Phys. **79**, 317 (1981)

[18] E. Corrigan, D. Olive, D. Fairlie et J. Nuyts, Nucl. Phys. **B106** , 475 (1976)

GENERALIZED LEVI-CIVITA CONNECTIONS

Roberto Percacci
International School for Advanced Studies, 34014 Trieste, Italy

Abstract. The fundamental theorem of riemannian geometry is generalized to the case of vectorbundles which are only partially soldered and have a possibly degenerate fiber metric.

1.Introduction

Einstein's General Relativity theory can be reformulated in such a way that its dynamical variables are defined in a vectorbundle ξ with fiber R^n which is isomorphic, but not canonically isomorphic, to the tangent bundle TM (M is spacetime and $n = dimM$). In this reformulation, the dynamical variables are an isomorphism θ of TM to ξ (the soldering form), a fiber metric κ in ξ and a linear connection ∇ in ξ. The local representatives of θ and κ in suitable bases have to be nondegenerate:

$$det(\theta^m{}_\mu(x)) \neq 0, \quad 1.1$$
$$det(\kappa_{mn}(x)) \neq 0; \quad 1.2$$

in addition, κ must have a fixed signature. The gauge group is the group of all linear automorphisms of ξ; either θ or κ can be "gauged away", recovering the standard "metric" or "n-bein" formulations, respectively [1,2]. Using a generalized Palatini formalism, the conditions that ∇ be torsionfree and metric:

$$d_\nabla \theta = 0, \quad 1.3$$
$$\nabla \kappa = 0, \quad 1.4$$

can be obtained as equations of motion, instead of being imposed as a priori constraints [1]. As is well known, given θ and κ satisfying 1.1 and 1.2, the equations 1.3 and 1.4 have a unique solution for ∇, the Levi-Civita connection.

In order to construct unified models of gravitational and Yang-Mills interactions, it is possible to generalize this theory by allowing ξ to have fiber R^N, with $N > n$ [3,4,5]. Furthermore, it appears likely that in the quantum theory both θ and κ have to be allowed to become degenerate. It is of obvious interest to study the solutions of equations 1.3 and 1.4 under these more general conditions. If ξ is any vectorbundle with fiber R^N, $N \geq n$, θ is a vectorbundle homomorphism of TM to ξ and κ is a quadratic form in the fibers of ξ, we call a connection ∇ in ξ satisfying 1.3 and 1.4 a Generalized Levi-Civita Connection

(GLCC). In sections 2 and 3 we solve completely the problem when θ, κ and $g = \theta^*\kappa$ have constant ranks: we give necessary and sufficient conditions on θ and κ for the existence of a GLCC and we describe the space of all GLCC's, when they exist. In section 4 we discuss briefly the case when θ, κ and g have nonconstant ranks; in section 5 we describe the space of gauge-inequivalent GLCC's.

The problem addressed in this work has been discussed previously in the literature in the special case $N = n$ and $\theta = Id_{TM}$ [6,7,8]; I am grateful to C. Isham and A. Trautman for calling my attention on these references. I also wish to thank P. Michor and G. Kainz for a useful discussion and for reading a first draft of this work.

2.Statement of the results

We work in the category C^∞. Let M be an n-dimensional manifold , ξ a real vectorbundle over M with fiber R^N, $N \geq n$. Let θ be a one-form on M with values in ξ, i.e. a section of $T^*M \otimes \xi$; since $T^*M \otimes \xi \approx Hom(TM, \xi)$,we may also regard θ as a vectorbundle homomorphism $TM \to \xi$. In particular, if θ has constant maximal rank n (i.e. it is a vectorbundle monomorphism), then it will be called a soldering form. Let κ be a symmetric bilinear form on ξ, i.e. a section of the symmetric tensorproduct $\xi^* \odot \xi^*$; in particular, if κ has constant maximal rank N, then it is a (pseudo) fiber-metric in ξ. Given a linear connection ∇ in ξ, we can form the covariant exterior derivative of θ, which is defined by

$$d_\nabla\theta(X,Y) = \nabla_X(\theta(Y)) - \nabla_Y(\theta(X)) - \theta([X,Y]) \qquad 2.1$$

for any couple of vectorfields $X, Y \in C^\infty(TM)$, and the covariant derivative of κ , which is defined by

$$(\nabla_X\kappa)(v,w) = X(\kappa(v,w)) - \kappa(\nabla_X v, w) - \kappa(v, \nabla_X w) \qquad 2.2$$

for any $X \in C^\infty(TM); v, w \in C^\infty(\xi)$. When (or where) θ has constant rank, there are defined subbundles $ker\theta$ of TM and $im\theta$ of ξ; we can choose complementary subbundles η of TM and ς of ξ, such that

$$TM = \eta \oplus ker\theta, \qquad 2.3$$

$$\xi = im\theta \oplus \varsigma. \qquad 2.4$$

We require that the vectors in ς are orthogonal to those in $im\theta$ with respect to the form κ; when κ is nondegenerate this completely specifies the splitting 2.4. θ provides an isomorphism of $\eta \subset TM$ to $im\theta \subset \xi$ and so may be regarded as a partial soldering. If η and ς are given, we can define an induced connection D in η by

$$D_X Y = (\theta^*\nabla)_X Y = (\theta|_\eta)^{-1}(\nabla_X(\theta(Y))|_{im\theta}) \qquad 2.5$$

for $X \in C^\infty(TM), Y \in C^\infty(\eta)$, and an η-valued two-form $\theta^*(d_\nabla\theta)$ by

$$\begin{aligned}(\theta^*(d_\nabla\theta))(X,Y) &= (\theta|_\eta)^{-1}(d_\nabla\theta(X,Y)|_{im\theta}) \\ &= D_XY - D_YX - [X,Y] = d_D(Id_\eta).\end{aligned} \tag{2.6}$$

In particular, if θ has maximal rank, D is a linear connection in TM and $\theta^*(d_\nabla\theta)$ is the torsion form of D in the usual sense. When (or where) κ has constant rank, there is defined a subbundle $\xi_0 = ker\kappa$ of ξ; we choose a complementary subbundle ξ_+ of ξ such that

$$\xi = \xi_+ \oplus \xi_0. \tag{2.7}$$

The form $\kappa|_{\xi_+}$ is nondegenerate, i.e. a true fiber metric; its signature will be irrelevant for our purposes. Now consider the forms $\kappa|_{im\theta}$ and $\kappa|_\varsigma$; when (or where) they have constant ranks, we can repeat what was said for κ and arrive at splittings

$$im\theta = im\theta_+ \oplus im\theta_0 \tag{2.8}$$

$$\varsigma = \varsigma_+ \oplus \varsigma_0. \tag{2.9}$$

It is clear that

$$\xi_0 = im\theta_0 \oplus \varsigma_0, \tag{2.10}$$

$$\xi_+ = im\theta_+ \oplus \varsigma_+. \tag{2.11}$$

Using $(\theta|_\eta)^{-1}$, 2.8 induces a splitting

$$\eta = \eta_+ \oplus \eta_0. \tag{2.12}$$

Given θ and κ we can define an induced symmetric bilinear form $g = \theta^*\kappa$ in TM. It is clear that

$$kerg = ker\theta \oplus \eta_0, \tag{2.13}$$

$$TM = \eta_+ \oplus kerg. \tag{2.14}$$

If we put $rank(\theta) = r$, $0 \le r \le n$; $rank(\kappa) = q$, $0 \le q \le N$; $rank(g) = q_1$, $0 \le q_1 \le min(q,r)$; $q = q_1 + q_2$, then the fiber dimensions of these vectorbundles are as follows:

$ker\theta$	η_0	η_+	ξ_0	ξ_+	$im\theta_0$	$im\theta_+$	ς_0	ς_+	ς
$n-r$	$r-q_1$	q_1	$N-q$	q	$r-q_1$	q_1	$N-r-q_2$	q_2	$N-r$

For each of these vectorbundles we define a corresponding projector Π; e.g. $\Pi_\eta : TM \to \eta$, $\Pi_\varsigma : \xi \to \varsigma$, a.s.o.

Recall that the space of all linear connections in ξ is an affine space modelled on $C^\infty(T^*M \otimes End\xi)$: if ∇_0 is any connection, every connection ∇ can be written uniquely $\nabla = \nabla_0 + \omega$, where ω is a one-form with values in $End\xi$. The splitting 2.11 induces a Z_2 grading in $End\xi_+$:

$$End\xi_+ = End_0\xi_+ \oplus End_1\xi_+,$$

where $End_0\xi_+ = End(im\theta_+) \oplus End\varsigma_+$ and $End_1\xi_+ = Hom(im\theta_+, \varsigma_+) \oplus Hom(\varsigma_+, im\theta_+)$. As in [1], given θ and κ we can define subbundles $T^*M \odot_\theta End\xi$ and $T^*M \otimes End^A\xi$: a belongs to the fiber over x of $T^*M \odot_\theta End\xi$ if and only if

$$a_X(\theta(Y)) = a_Y(\theta(X)) \qquad \forall\, X, Y \in TM_x, \tag{2.15}$$

and a belongs to the fiber over x of $T^*M \otimes End^A\xi$ if and only if

$$\kappa(a_X(v), w) + \kappa(v, a_X(w)) = 0 \qquad \forall\, X \in TM_x; v, w \in \xi_x. \tag{2.16}$$

Here $a_X = a(X)$ is an element of $End\xi_x$. The same type of restrictions can be put on subbundles of $T^*M \otimes End\xi$. In particular, we will need the subbundles $\eta_+^* \odot_\theta End_1^A\xi_+$, $T^*M \otimes End^A\varsigma_+$, $T^*M \otimes Hom(\varsigma, \xi_0)$ and $\eta^* \odot_\theta Hom(im\theta, \xi_0)$ which are defined as follows (we omit the subscripts x to denote the fiber, for notational simplicity):

$$a \in \eta_+^* \odot_\theta End_1^A\xi_+ \iff X \in ker g \Rightarrow a_X = 0 \tag{2.17}$$
$$v \in \xi_0 \Rightarrow a_X(v) = 0 \tag{2.18}$$
$$v \in im\theta_+ \Rightarrow a_X(v) \in \varsigma_+ \tag{2.19}$$
$$v \in \varsigma_+ \Rightarrow a_X(v) \in im\theta_+ \tag{2.20}$$
$$a_X(\theta(Y)) = a_Y(\theta(X)) \tag{2.21}$$
$$\kappa(a_X(v), w) = -\kappa(v, a_X(w)) \tag{2.22}$$
$$a \in T^*M \otimes End^A\varsigma_+ \iff v \in im\theta \oplus \varsigma_0 \Rightarrow a_X(v) = 0 \tag{2.23}$$
$$v \in \varsigma_+ \Rightarrow a_X(v) \in \varsigma_+ \tag{2.24}$$
$$\kappa(a_X(v), w) = -\kappa(v, a_X(w)) \tag{2.25}$$
$$a \in T^*M \otimes Hom(\varsigma, \xi_0) \iff v \in im\theta \Rightarrow a_X(v) = 0 \tag{2.26}$$
$$v \in \varsigma \Rightarrow a_X(v) \in \xi_0 \tag{2.27}$$
$$a \in \eta^* \odot_\theta Hom(im\theta, \xi_0) \iff X \in ker\theta \Rightarrow a_X = 0 \tag{2.28}$$
$$v \in \varsigma \Rightarrow a_X(v) = 0 \tag{2.29}$$
$$v \in im\theta \Rightarrow a_X(v) \in \varsigma_0 \tag{2.30}$$
$$a_X(\theta(Y)) = a_Y(\theta(X)) \tag{2.31}$$

We are now ready to state the main result of this work.

Theorem. **Suppose θ, κ and g are globally defined on M and have constant ranks r, q and q_1 respectively. Then:**
a) a GLCC exists if and only if the following three conditions are satisfied: i) the distribution on M defined by $kerg$ is integrable, so M is foliated with $(n - q_1)$-dimensional leaves; ii) the distribution $ker\theta \subset kerg$ is integrable, so the leaves of the foliation defined by $kerg$ are in turn foliated with $(n - r)$-dimensional leaves; iii) the Lie derivative of g along the leaves of the foliation defined by $ker\theta$ is zero;
b) if ∇_0 is a GLCC, then every other GLCC can be written uniquely in the form $\nabla = \nabla_0 + \alpha + \beta + \gamma + \delta$ where $\alpha \in C^\infty(\eta_+^* \odot_\theta End_1^A \xi_+)$, $\beta \in C^\infty(T^*M \otimes End^A \varsigma_+)$, $\gamma \in C^\infty(T^*M \otimes Hom(\varsigma, \xi_0))$ and $\delta \in C^\infty(\eta^* \odot_\theta Hom(im\theta, \xi_0))$.

Half of the proof of this theorem will be given here using coordinate-independent methods. We first prove the necessity of the conditions in a), i.e. we show that if equations 1.3 and 1.4 hold, then conditions i), ii), and iii) are satisfied.

By Frobenius' theorem, i) is equivalent to the statement that

$$X, Y \in C^\infty(kerg) \quad \Rightarrow \quad [X, Y] \in C^\infty(kerg) \qquad 2.32$$

and ii) is equivalent to the statement that

$$X, Y \in C^\infty(ker\theta) \quad \Rightarrow \quad [X, Y] \in C^\infty(ker\theta). \qquad 2.33$$

Because of 2.14, condition iii) is equivalent to the statement that if $Z \in C^\infty(ker\theta)$ and $X, Y \in C^\infty(\eta_+)$,

$$Z(g(X, Y)) = g([Z, X], Y) + g(X, [Z, Y]). \qquad 2.34$$

To prove 2.33, let $X, Y \in C^\infty(ker\theta)$; 1.3 implies $0 = d_\nabla \theta(X, Y) = -\theta([X, Y])$, so $[X, Y] \in C^\infty(ker\theta)$. To prove 2.32, consider first the case $X \in C^\infty(\eta_0)$, $Y \in C^\infty(ker\theta)$; we have $d_\nabla \theta(X, Y) = -\nabla_Y(\theta(X)) - \theta([X, Y])$ so 1.3 implies in particular $\nabla_Y(\theta(X))|_{im\theta_+} + \theta([X, Y])|_{im\theta_+} = 0$. If $Z \in C^\infty(\eta_+)$, 1.4 implies $\kappa(\nabla_Y(\theta(X)), \theta(Z)) = 0$; since $\theta(Z)$ is an arbitrary section of $im\theta_+$ and κ is nondegenerate on $im\theta_+$, $\nabla_Y(\theta(X))|_{im\theta_+} = 0$. So $\theta([X, Y])|_{im\theta_+} = 0$, or $X, Y \in kerg$. Next consider the case $X, Y \in C^\infty(\eta_0)$; 1.3 implies $\nabla_X(\theta(Y))|_{im\theta_+} - \nabla_Y(\theta(X))|_{im\theta_+} = \theta([X, Y])|_{im\theta_+}$. If $Z \in C^\infty(\eta_+)$, 1.4 implies that $\kappa(\nabla_Y(\theta(X)), \theta(Z)) = 0$ and $\kappa(\nabla_X(\theta(Y)), \theta(Z)) = 0$, and since $\theta(Z) \in C^\infty(\eta_+)$ is arbitrary, $\nabla_Y(\theta(X))|_{im\theta_+} = \nabla_X(\theta(Y))|_{im\theta_+} = 0$. So, again, $\theta([X, Y])|_{im\theta_+} = 0$ and $[X, Y] \in kerg$. These two results, together with 2.33 and 2.13, prove 2.32. To prove 2.34 we note that if $X, Y \in C^\infty(\eta_+)$ and $Z \in$

$C^\infty(ker\theta)$, 1.3 implies that $\nabla_Z(\theta(X)) = \theta([Z,X])$ and $\nabla_Z(\theta(Y)) = \theta([Z,Y])$. Therefore, 1.4 implies

$$\begin{aligned} 0 &= (\nabla_Z\kappa)(\theta(X),\theta(Y)) \\ &= Z(\kappa(\theta(X),\theta(Y))) - \kappa(\nabla_Z(\theta(X)),\theta(Y)) - \kappa(\theta(X),\nabla_Z(\theta(Y))) \\ &= Z(g(X,Y)) - g([Z,X],Y) - g(X,[Z,Y]). \end{aligned}$$

We now show that if ∇_0 is a GLCC, also $\nabla = \nabla_0 + \alpha + \beta + \gamma + \delta$ is a GLCC. We have

$$\begin{aligned} d_\nabla\theta(X,Y) = \ &\alpha_X(\theta(Y)) + \beta_X(\theta(Y)) + \gamma_X(\theta(Y)) + \delta_X(\theta(Y)) \\ &- \alpha_Y(\theta(X)) - \beta_Y(\theta(X)) - \gamma_Y(\theta(X)) - \delta_Y(\theta(X)). \end{aligned}$$

It is sufficient to consider this equation at a single point. If both $X, Y \in ker\theta$, both lines are trivially zero. If $X \in ker\theta$ and $Y \in \eta$, the second line vanishes trivially and in the first line $\alpha_X(\theta(Y)) = 0$ by 2.17, $\beta_X(\theta(Y)) = 0$ by 2.23, $\gamma_X(\theta(Y)) = 0$ by 2.26 and $\delta_X(\theta(Y)) = 0$ by 2.28. If both $X, Y \in \eta$, $\beta_X(\theta(Y)) = \beta_Y(\theta(X)) = 0$ by 2.23, $\gamma_X(\theta(Y)) = \gamma_Y(\theta(X)) = 0$ by 2.26, $\delta_X(\theta(Y)) - \delta_Y(\theta(X)) = 0$ by 2.31; for α we have to further distinguish: if $X, Y \in \eta_+$, $\alpha_X(\theta(Y)) - \alpha_Y(\theta(X)) = 0$ by 2.21, if $X \in \eta_+$, $Y \in \eta_0$, $\alpha_X(\theta(Y)) = 0$ by 2.18 and $\alpha_Y(\theta(X)) = 0$ by 2.17, if $X, Y \in \eta_0$, $\alpha_X(\theta(Y)) = \alpha_Y(\theta(X)) = 0$ by 2.17. So for any choice of $X, Y \in TM$, $d_\nabla\theta(X,Y) = 0$. Next, we have

$$\begin{aligned} (\nabla_X\kappa)(v,w) = &- \kappa(\alpha_X(v),w) - \kappa(\beta_X(v),w) - \kappa(\gamma_X(v),w) - \kappa(\delta_X(v),w) \\ &- \kappa(v,\alpha_X(w)) - \kappa(v,\beta_X(w)) - \kappa(v,\gamma_X(w)) - \kappa(v,\delta_X(w)). \end{aligned}$$

Consider first the terms containing α; by 2.17 they vanish when $X \in kerg$, so assume $X \in \eta_+$. If $v \in \xi_0$, $\kappa(v,\alpha_X(w)) = 0$ by definition and $\alpha_X(v) = 0$ by 2.18; similarly if $w \in \xi_0$. If both $v, w \in \xi_+$, $\kappa(\alpha_X(v),w) + \kappa(v,\alpha_X(w)) = 0$ by 2.22. Now consider the terms containing β. If both $v, w \in im\theta \oplus \varsigma_0$, $\beta_X(v) = \beta_X(w) = 0$ by 2.23; if $v \in im\theta \oplus \varsigma_0, w \in \varsigma_+$, $\beta_X(v) = 0$ by 2.23 and $\kappa(v,\beta_X(w)) = 0$ by 2.24; if both $v, w \in \varsigma_+$, $\kappa(\beta_X(v),w) + \kappa(v,\beta_X(w)) = 0$ by 2.25. Now consider the terms containing γ. If both $v, w \in im\theta$, $\gamma_X(v) = \gamma_X(w) = 0$ by 2.26; if $v \in im\theta, w \in \varsigma$, $\gamma_X(v) = 0$ by 2.26 and $\kappa(v,\gamma_X(w)) = 0$ by 2.27; if both $v, w \in \varsigma$, $\kappa(\gamma_X(v),w) = \kappa(v,\gamma_X(w)) = 0$ by 2.27. Finally consider the terms containing δ. If $X \in ker\theta$, $\delta_X(v) = \delta_X(w) = 0$ by 2.28, so assume $X \in \eta$. If both $v, w \in \varsigma$, $\delta_X(v) = \delta_X(w) = 0$ by 2.29; if $v \in \varsigma, w \in im\theta$, $\delta_X(v) = 0$ by 2.29 and $\kappa(v,\delta_X(w)) = 0$ by 2.30; if both $v, w \in im\theta$, $\kappa(\delta_X(v),w) = \kappa(v,\delta_X(w)) = 0$ by 2.31. So for any choice of X, v, w, $(\nabla_X\kappa)(v,w) = 0$.

It remains to be proven that the conditions given in a) are also sufficient for the existence of a GLCC, and that *every* GLCC can be obtained from ∇_0 by adding the forms α, β, γ and δ. In order to do this, we shall first solve the problem in local coordinates. We shall conclude the proof of the theorem at the end of section 3.

3. Local results.

Let us introduce the following indexing sets: $I_{\eta+} = \{1,\dots,q_1\}$, $I_{\eta_0} = \{q_1+1,\dots,r\}$, $I_{ker\theta} = \{r+1,\dots,n\}$, $I_\eta = I_{\eta+} \cup I_{\eta_0}$, $I_{TM} = I_\eta \cup I_{ker\theta}$, $I_{kerg} = I_{\eta_0} \cup I_{ker\theta}$, $I_{im\theta_+} = \{1,\dots,q_1\}$, $I_{im\theta_0} = \{q_1+1,\dots,r\}$, $I_{\varsigma+} = \{r+1,\dots,r+q_2\}$, $I_{\varsigma_0} = \{r+q_2+1,\dots,N\}$, $I_{im\theta} = I_{im\theta_+} \cup I_{im\theta_0}$, $I_\varsigma = I_{\varsigma+} \cup I_{\varsigma_0}$, $I_{\xi+} = I_{im\theta_+} \cup I_{\varsigma+}$, $I_{\xi_0} = I_{im\theta_0} \cup I_{\varsigma_0}$, $I_\xi = I_{im\theta} \cup I_\varsigma = I_{\xi+} \cup I_{\xi_0}$.

Let U be an open subset of M such that $TM|_U$ and $\xi|_U$ are trivializable, and suppose θ, κ and g have constant ranks r, q and q_1 on U respectively; we introduce n linearly independent sections $\{L_\mu\}_{\mu\in I_{TM}}$ of $TM|_U$ and N linearly independent sections $\{e_m\}_{m\in I_\xi}$ of $\xi|_U$, such that when the indices μ and m run over a subset of I_{TM} and I_ξ , the sections labelled by those indices span the corresponding subbundle (e.g. $\{L_\mu\}_{\mu\in I_\eta}$ are q linearly independent sections of $\eta|_U$). Let us define the following local components:

$$\theta^m{}_\mu = \langle e^m|\theta(L_\mu)\rangle \tag{3.1}$$

$$\kappa_{mn} = \kappa(e_m, e_n) \tag{3.2}$$

$$\Gamma_\lambda{}^m{}_n = \langle e^m|\nabla_{L_\mu} e_n\rangle \tag{3.3}$$

$$c^\lambda{}_{\rho\sigma} = \langle L^\lambda|[L_\rho, L_\sigma]\rangle \tag{3.4}$$

where $\{e^m\}_{m\in I_\xi}$ and $\{L^\mu\}_{\mu\in I_{TM}}$ are the dual bases in ξ^* and T^*M. 1.3 is equivalent to the following set of $\frac{1}{2}Nn(n-1)$ equations

$$L_\mu(\theta^m{}_\nu) - L_\nu(\theta^m{}_\mu) + \Gamma_\mu{}^m{}_n\theta^n{}_\nu - \Gamma_\nu{}^m{}_n\theta^n{}_\mu - \theta^m{}_\rho c^\rho{}_{\mu\nu} = 0 \qquad (\nu<\mu) \tag{3.5}$$

and 1.4 is equivalent to the following $\frac{1}{2}Nn(N+1)$ equations

$$L_\mu(\kappa_{mn}) - \Gamma_\mu{}^l{}_m\,\kappa_{ln} - \Gamma_\mu{}^l{}_n\,\kappa_{ml} = 0 \qquad (n\le m). \tag{3.6}$$

For fixed θ and κ, we regard 3.5 and 3.6 as an infinite system of linear equations ($\frac{1}{2}Nn(N+n)$ equations per spacetime point) for the connection coefficients $\Gamma_\lambda{}^m{}_n$. Actually, we may collect the connection coefficients into a column vector $\Gamma \in C^\infty(U, R^{nN^2})$ with multi-index $\lfloor_\lambda{}^m{}_n\rfloor$ and write 3.5, 3.6 in the form

$$A(x)\Gamma(x) = B(x) \tag{3.7}$$

where $B \in C^\infty(U, R^{\frac{1}{2}Nn(N+n)})$ and $A \in C^\infty(U, Hom(R^{nN^2}, R^{\frac{1}{2}Nn(N+n)}))$ are defined by

$$\begin{aligned}
B[^m{}_{\mu\nu}] &= \theta^m{}_\rho c^\rho{}_{\mu\nu} - L_\mu(\theta^m{}_\nu) + L_\nu(\theta^m{}_\mu) && (\nu<\mu)\\
B[_{\mu mn}] &= L_\mu(\kappa_{mn}) && (n\le m)\\
A[^m{}_{\mu\nu}][^\lambda{}_r{}^s] &= \delta^m_r(\delta^\lambda_\mu\theta^s{}_\nu - \delta^\lambda_\nu\theta^s{}_\mu) && (\nu<\mu)\\
A[_{\mu mn}][^\lambda{}_r{}^s] &= \delta^\lambda_\mu(\kappa_{mr}\delta^s_n + \kappa_{rn}\delta^s_m) && (n\le m).
\end{aligned}$$

Applying for each $x \in U$ the usual theory of linear equations, we get the following: 3.7 has a solution if and only if $B(x) \in imA(x) \quad \forall x \in U$ and the space of independent solutions is $A^{-1}(B) \approx C^\infty(U, R^Q)$ with $Q = nN^2 - rankA$. We want now to determine the number Q and, more precisely, what components of Γ are left undetermined by equations 3.5, 3.6. We will do this by direct inspection and counting.

Proposition. Suppose $\theta^m{}_\mu$, κ_{mn} and $g_{\mu\nu}$ have constant ranks r, q and q_1 respectively on U. Then:
a) the system 3.5, 3.6 has a solution if and only if the following conditions hold: i)$c^\lambda{}_{\mu\nu} = 0$ for $\lambda \in I_{\eta+}$, $\mu, \nu \in I_{kerg}$; ii) $c^\lambda{}_{\mu\nu} = 0$ for $\lambda \in I_\eta$, $\mu, \nu \in I_{ker\theta}$; iii)$L_\lambda(g_{\mu\nu}) = g_{\mu\tau}c^\tau{}_{\lambda\nu} + g_{\nu\tau}c^\tau{}_{\lambda\mu}$ for $\lambda \in I_{ker\theta}$, $\mu, \nu \in I_{\eta+}$;
b) there is a bijective correspondence between the set of solutions of the system 3.5, 3.6 over U and $C^\infty(U, R^Q)$ where $Q = \frac{1}{2}q_2q_1(q_1 + 1) + \frac{1}{2}nq_2(q_2 - 1) + (N - q)[n(N - r) + \frac{1}{2}r(r + 1)]$; the independent free components of $\Gamma_\lambda{}^m{}_n$ can be chosen as follows: α) $\lambda \in I_{\eta+}$, $m \in I_{\varsigma+}$, $n \in I_{im\theta+}$, $\lambda \geq n$; β) $\lambda \in I_{TM}$, $m, n \in I_{\varsigma+}$, $m > n$; γ) $\lambda \in I_\eta$, $m \in I_{\xi_0}$, $n \in I_{im\theta}$, $\lambda \geq n$; δ) $\lambda \in I_{TM}$, $m \in I_{\xi_0}$, $n \in I_\varsigma$.

Proof. It will be convenient to choose the bases in a particular way. Since $\theta|_\eta$ is an isomorphism, we demand that $e_\mu = \theta(L_\mu)$ for $\mu \in I_\eta = I_{im\theta}$, so that the matrix $\theta^m{}_\mu$ has the form $\theta^m{}_\mu = \delta^m_\mu$ for $m, \mu \in I_\eta = I_{im\theta}$ and $\theta^m{}_\mu = 0$ otherwise. Then, by means of a $GL(q)$-transformation in ξ_+, followed by an appropriate $GL(q_1)$- transformation in η_+ in order to preserve the form of $\theta^m{}_\mu$, we go to bases in which the matrix κ_{mn} is diagonal with eigenvalues $\lambda_m = \pm 1$ for $m \in I_{\xi_+}$ and $\lambda_m = 0$ for $m \in I_{\xi_0}$, and the matrix $g_{\mu\nu}$ is diagonal with eigenvalues $\lambda_\mu = \pm 1$ for $\mu \in I_{\eta_0}$ and $\lambda_\mu = 0$ for $\mu \in I_{kerg}$. In these bases, the system 3.5, 3.6 can be decomposed as follows:

$$m \in I_\eta = I_{im\theta}; \mu, \nu \in I_\eta = I_{im\theta}: \quad \Gamma_\mu{}^m{}_\nu - \Gamma_\nu{}^m{}_\mu = c^m{}_{\mu\nu}; \qquad 3.8$$

$$m \in I_\eta = I_{im\theta}; \mu \in I_\eta = I_{im\theta}; \nu \in I_{ker\theta}: \quad \Gamma_\nu{}^m{}_\mu + c^m{}_{\mu\nu} = 0; \qquad 3.9a$$

$$m \in I_\eta = I_{im\theta}; \mu \in I_{ker\theta}; \nu \in I_\eta = I_{im\theta}: \quad \Gamma_\mu{}^m{}_\nu - c^m{}_{\mu\nu} = 0; \qquad 3.9b$$

$$m \in I_\eta = I_{im\theta}; \mu, \nu \in I_{ker\theta}: \quad c^m{}_{\mu\nu} = 0; \qquad 3.10$$

$$m \in I_\varsigma; \mu, \nu \in I_\eta = I_{im\theta}: \quad \Gamma_\mu{}^m{}_\nu - \Gamma_\nu{}^m{}_\mu = 0; \qquad 3.11$$

$$m \in I_\varsigma; \mu \in I_\eta = I_{im\theta}; \nu \in I_{ker\theta}: \quad \Gamma_\nu{}^m{}_\mu = 0; \qquad 3.12a$$

$$m \in I_\varsigma; \mu \in I_{ker\theta}; \nu \in I_\eta = I_{im\theta}: \quad \Gamma_\mu{}^m{}_\nu = 0; \qquad 3.12b$$

$$m \in I_\varsigma; \mu, \nu \in I_{ker\theta}: \quad 0 = 0;$$

$$\mu \in I_{TM}; m, n \in I_{\xi_+}: \quad \lambda_n \Gamma_\mu{}^n{}_m + \lambda_m \Gamma_\mu{}^m{}_n = 0; \quad 3.13$$

$$\mu \in I_{TM}; m \in I_{\xi_+}; n \in I_{\xi_0}: \quad \Gamma_\mu{}^m{}_n = 0; \qquad 3.14a$$

$$\mu \in I_{TM}; m \in I_{\xi_0}; n \in I_{\xi_+}: \quad \Gamma_\mu{}^n{}_m = 0; \qquad 3.14b$$

$$\mu \in I_{TM}; m, n \in I_{\xi_0}: \quad 0 = 0.$$

No summation over repeated indices is implied in 3.13. Notice that the equations 3.9a,b; 3.12a,b and 3.14a,b are pairwise equivalent. Condition i) follows from 3.10 when $\mu,\nu \in I_{ker\theta}$, from 3.9 and 3.14 when $\mu \in I_{ker\theta}$ and $\nu \in I_{\eta_0}$ or vice-versa, and from 3.8 and 3.14 when both $\mu,\nu \in I_{\eta_0}$; condition ii) is equivalent to 3.10; condition iii) follows from 3.9 and 3.13.

We arrange the components $\Gamma_\lambda{}^m{}_n$ into an $n \times N \times N$ parallelepiped as in fig.1.

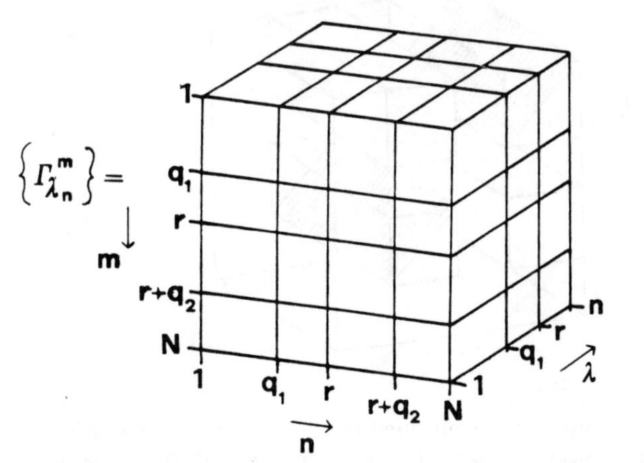

Fig.1

3.8 and 3.11 are relations between the components contained in the $r\times N\times r$ prism shown in fig.2: the components on one side of the dotted diagonal plane are determined by those on the other side. Altogether 3.8 and 3.11 amount to $\frac{1}{2}Nr(r-1)$ independent equations.

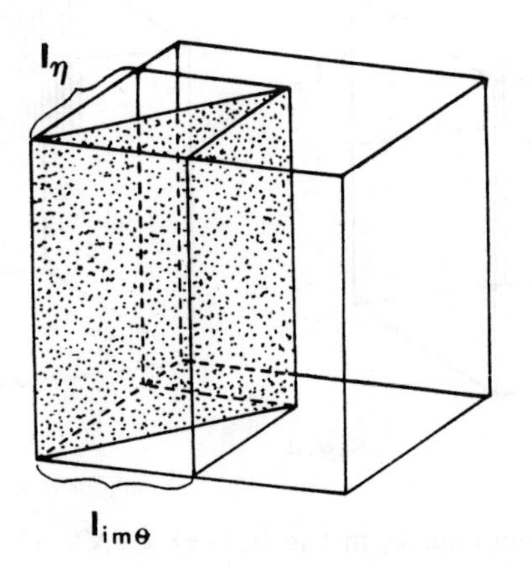

Fig.2

3.13 are relations between the coefficients contained in the four boxes drawn in fig.3: the coefficients on one side of the striped plane are determined by those on the other side, and those on the plane vanish.They give $\frac{1}{2}nq(q+1)$ independent equations.

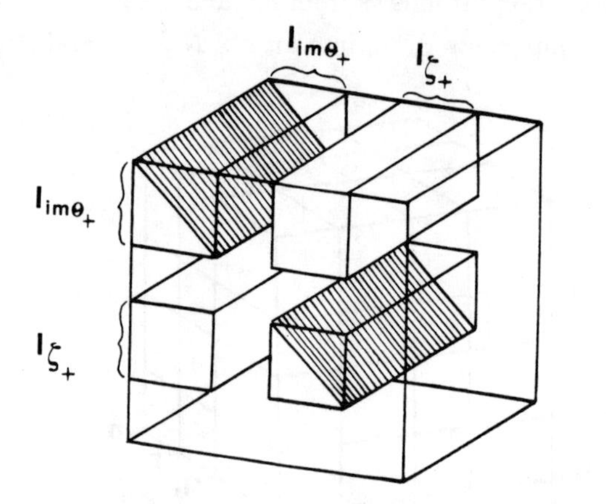

Fig.3

3.9 fixes the components contained in the $(n-r)\times r\times r$ prism shown on the right in fig.4. However due to 3.13 not all these relations are independent: the independent ones can be chosen as shown on the left, and their number is $(n-r)[r^2 - \frac{1}{2}q_1(q_1+1)]$.

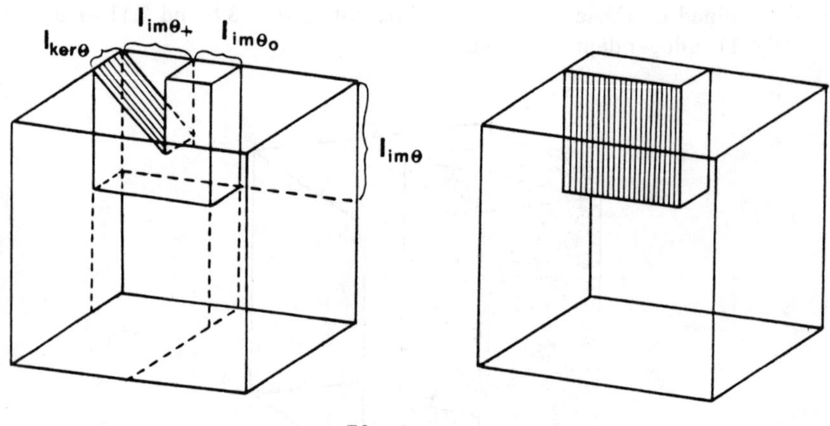

Fig.4

3.12 forces the components in the $(n-r)\times(N-r)\times r$ prism shown on the left in fig.5 to vanish. All these relations are independent. Due to 3.13 also other components, contained in a $(n-r) \times q_1 \times q_2$ prism, will have to vanish. All the components which are determined in this way are shown on the right in fig.5.

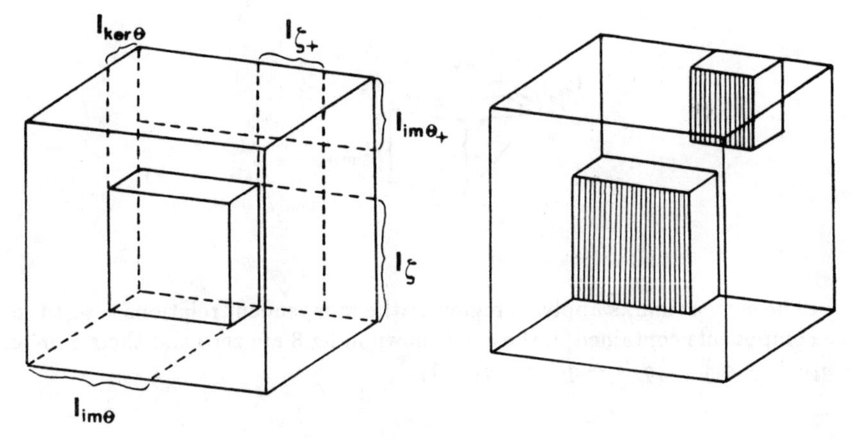

Fig.5

3.14 say that the components contained in the four prisms numbered 1,2,3,4 in fig.6 are zero. Due to 3.8, 3.11, 3.13, the components contained in three more boxes are forced to vanish; all the components which are determined in this way are shown on the right in fig.6.

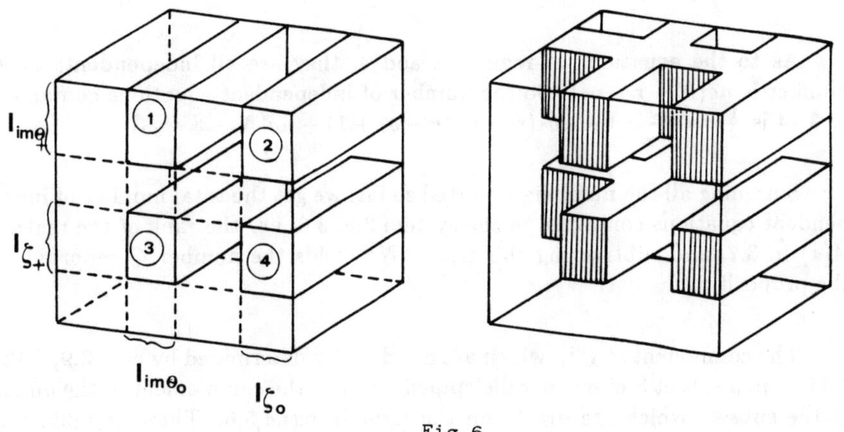

Fig.6

In order to count the number of independent equations contained in 3.14, we proceed separately in regions 1,2,3,4. In region 1, the components with $r+1 \leq \lambda \leq n$ were already fixed by 3.9, and those remaining are subject to the "symmetries" with respect to the diagonal planes of figs.2,3. The independent ones can be chosen as shown in fig.7 and their number is $\frac{1}{2}q_1(q_1-1)(r-q_1)+\frac{1}{2}q_1(r-q_1)(r-q_1+1)=\frac{1}{2}rq_1(r-q_1)$.

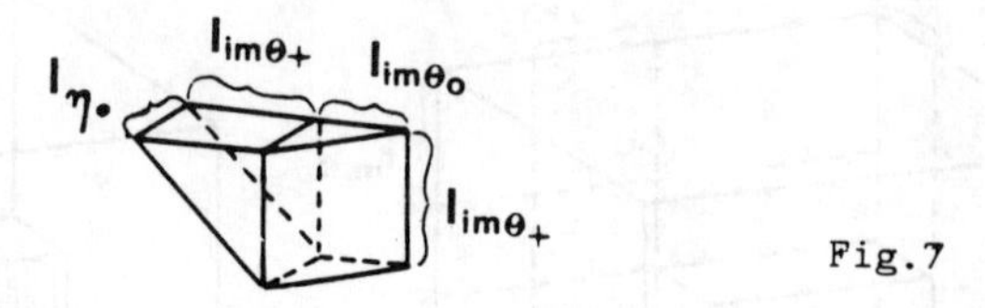

Fig.7

The same remarks apply to region 3: the independent relations assert that the components contained in the solid shown in fig.8 are zero and their number is $q_1q_2(r-q_1)+\frac{1}{2}q_2(r-q_1)(r-q_1+1)$.

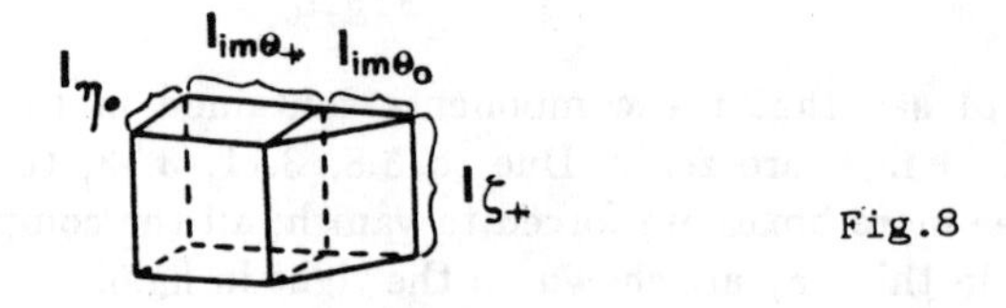

Fig.8

As to the equations of regions 2 and 4, they are all independent; their number is $nq(N-r-q_2)$. So the number of independent equations contained in 3.14 is $\frac{1}{2}rq_1(r-q_1)+\frac{1}{2}q_2(r-q_1)(r+q_1+1)+nq(N-r-q_2)$.

Summing all the numbers reported so far, we get the total number of independent equations contained in the system 3.5, 3.6, i.e. the rank of the matrix $A(x)$ in 3.7, and subtracting this from nN^2 yields the number Q reported in the proposition.

The components $\Gamma_\lambda{}^m{}_n$ which are not directly determined by eqs. 3.9, 3.12, 3.14 lie in a subset S of our parallelepiped which is the complement of the union of the subsets which are drawn on the right in figs.4,5,6. These components are related by eqs. 3.8, 3.11, 3.13, i.e. by the symmetries with respect to the diagonal planes of figs.2,3. We are now going to count how many of them are independent. To do this, we split S into the union of four subsets, which are shown on the right in figs.9,10,11,12.

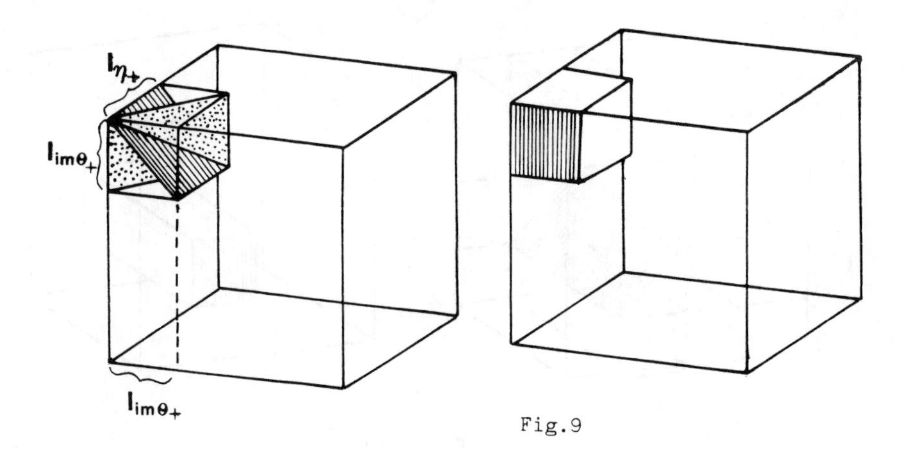

Fig.9

In the upper forward left cube of side q_1 there are q_1^3 components which are subject to q_1^3 independent equations: $\frac{1}{2}q_1^2(q_1-1)$ equations coming from 3.8 which are visualized as symmetry with respect to the dotted plane on the left in fig.9, and $\frac{1}{2}q_1^2(q_1+1)$ equations coming from 3.13 which are visualized as antisymmetry with respect to the striped plane. These are just the components of the Levi-Civita connection in the bundle η_+ determined by the soldering form Id_{η_+} and the fiber metric $g|_{\eta_+}$. So there are no free components in this cube.

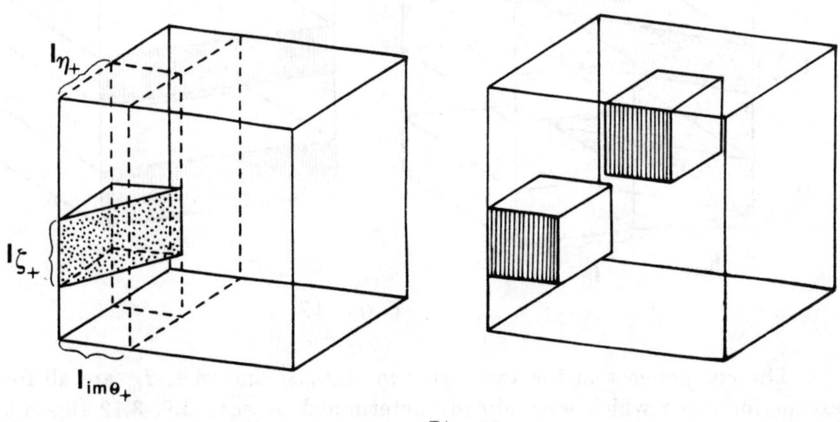

Fig.10

The components in the two prisms on the right in fig.10 are related by the symmetries of figs.2,3; the independent ones can be chosen as shown on the left and their number is $\frac{1}{2}q_1q_2(q_1+1)$.

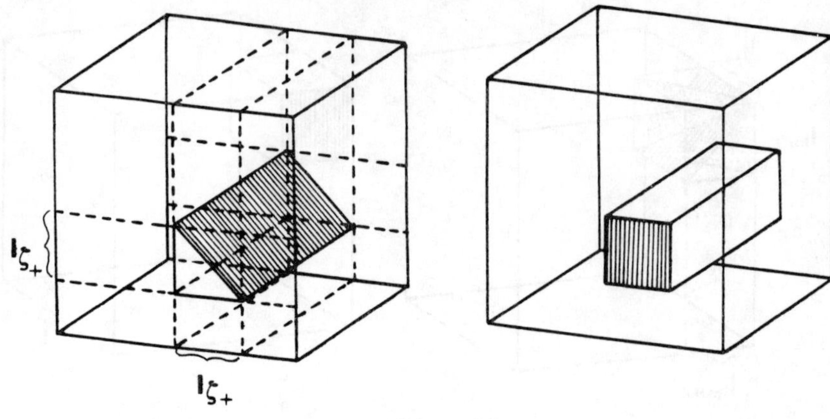

Fig. 11

The components in the prism on the right in fig.11 are related by the symmetry of fig.2. The independent ones are $\frac{1}{2}nq_2(q_2-1)$, as shown on the left.

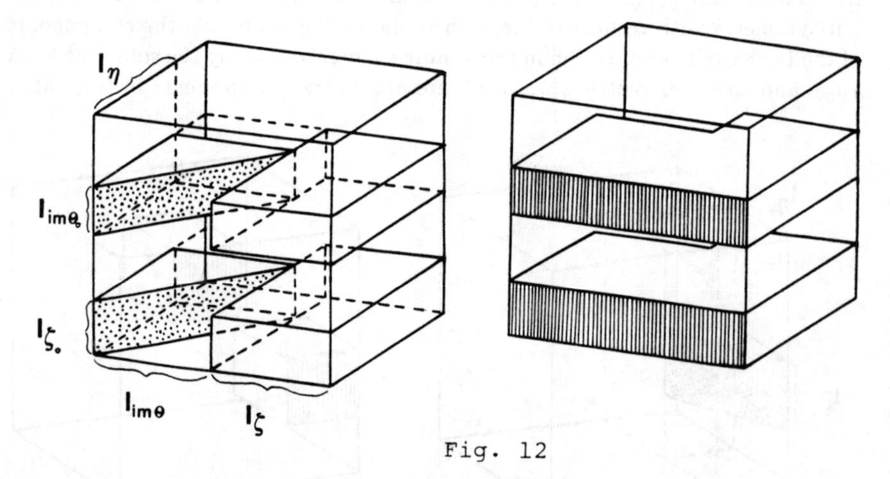

Fig. 12

The components in the two layers $m \in I_{im\theta_0}$ and $m \in I_{\zeta 0}$ are all free, except for those which were already determined by eqs. 3.9, 3.12 (figs.4,5), and for the symmetry with respect to the striped plane in fig.2. In fig.12 on the left is shown a choice of the independent components; their number is $(N-q)[n(N-r)+\frac{1}{2}r(r+1)]$.

It is now clear that the conditions listed in part a) of the proposition are necessary and sufficient for the existence of a solution of eqs. 3.5, 3.6 on U.

Part b) of the proposition follows from an inspection of figs.10,11,12. This concludes the proof of the proposition.

We can now conclude the proof of the theorem in section 2. First we observe that via 2.32, 2.33 and 2.34, the conditions i), ii) and iii) of the theorem are equivalent to the conditions i), ii) and iii) of the proposition. We have seen that these conditions are sufficient for the existence of a GLCC locally. To prove that a GLCC exists globally, let $\{U_A\}$ be an open cover of M which trivializes TM and ξ and let $\{f_A\}$ be a partition of unity relative to this cover. Let ∇_A be the unique local GLCC on U_A which is obtained by setting equal to zero all the arbitrary components which are listed in part b) of the proposition. Then $\nabla_0 = \sum_A f_A \nabla_A$ is a globally defined GLCC.

It remains to be proven that every GLCC can be obtained from ∇_0 by adding the forms α, β, γ and δ. This follows from part b) of the proposition, once we notice that with our choice of ∇_0, we can identify the independent components ${\Gamma_\lambda}^m{}_n$ listed under α), β), γ) and δ) with the components of the forms α, β, γ and δ.

4. Some topological remarks.

In the previous sections it was assumed that θ, κ and g have constant ranks r, q, and q_1 respectively. Due to the existence of topological obstructions, for a given bundle ξ there might not exist a θ and a κ of given, constant, positive ranks. Nevertheless, we shall show now that the discussion of the previous sections is valid for "almost all" θ's and κ's and on "almost all" of M.

Let us consider first the case of θ. Denote $L(n, N) = Hom(R^n, R^N)$, $L_r(n, N)$ the submanifold of linear maps of rank r, $L_{\leq r}(n, N) = \cup_{k=0}^r L_k(n, N)$ for $0 \leq r \leq n$. The interior of $L_{\leq r}(n, N)$ is $L_r(n, N)$ and its boundary is $L_{\leq n-1}(n, N)$; in particular $L_n(n, N)$ is open and dense in $L(n, N)$, while $L_{\leq n-1}(n, N)$ is nowhere dense. If $r < n$, $L_{\leq r}(n, N)$ is a submanifold with corners of $L(n, N)$ of codimension $(n-r)(N-r)$. This situation repeats itself fiberwise in the vectorbundle $Hom(TM, \xi)$, whose typical fiber is $L(n, N)$. Denote $Hom_r(TM, \xi)$ and $Hom_{\leq r}(TM, \xi)$ the subbundles with fibers $L_r(n, N)$ and $L_{\leq r}(n, N)$. $Hom_n(TM, \xi)$ is an open dense subbundle of $Hom(TM, \xi)$ and for each $0 \leq r \leq n-1$, $Hom_{\leq r}(TM, \xi)$ is a closed subbundle of codimension $(n-r)(N-r)$. Clearly, a section $\theta \in C^\infty(Hom(TM, \xi))$ has constant rank r if and only if it is a section of $Hom_r(TM, \xi)$. In view of these facts, a generic local section will have constant maximal rank n, but if we try to extend it to a global section of $Hom_n(TM, \xi)$ we might meet topological obstructions. The fiber of $Hom_n(TM, \xi)$ may be identified with the Stiefel manifold $V_n(R^N) = GL(N)/GL(N-n)$ of n-frames in R^N, because every n-frame in R^N defines a unique linear map $R^n \to R^N$, and vice-versa. The first $(N-n-1)$

homotopy groups of $V_n(R^N)$ vanish and so standard obstruction theory tells us that if $n \leq N - n$ there are no obstructions to finding global sections [9]. Thus, if ξ has fiber dimension $N > 2n - 1$, there exists θ of constant maximal rank n. If the fiber dimension of ξ is lower, there may or may not exist obstructions, depending on the topologies of M and ξ. We shall not go deeper into these questions.

A similar discussion could be carried out for κ, replacing $L(n, N)$ by $L(N, N)$ and $Hom(TM, \xi)$ by $\xi^* \odot \xi^*$. If κ has rank q, it defines a restriction of the structure group of ξ from $GL(N)$ to the subgroup of matrices of the form

$$\begin{bmatrix} A & 0 \\ B & C \end{bmatrix} \tag{4.1}$$

with $A \in O(a, b)$ with $a + b = q$, $C \in GL(N - q)$ and $B \in L(q, N - q)$. This subgroup is retractable to $O(a) \times O(b) \times O(N - q)$. Two types of obstructions could arise: the first could prevent ξ from being split as a direct sum of two vectorbundles, as in 2.8; the second could forbid the existence of a quadratic form of signature (a, b) in ξ_+. The reader is referred to the literature for more details [9].

Now we discuss the structure of the set where $rank(\theta)$ is not maximal. Denote $M_r^\theta = \theta^{-1}(Hom_r(TM, \xi))$ and $M_{\leq r}^\theta = \theta^{-1}(Hom_{\leq r}(TM, \xi))$ the subsets of M where θ has rank r and $\leq r$ respectively. Denote $\bar{r} = max_{x \in M}(rank\theta(x))$ so that θ can be regarded as a section of $Hom_{\leq \bar{r}}(TM, \xi)$; since θ is continuous, $M_{\bar{r}}^\theta$ is open and $M_{\leq r}^\theta$, for $0 \leq r \leq \bar{r} - 1$ are closed in M. It is not possible to say much more without imposing conditions on θ.

We say that a section $\theta \in C^\infty(Hom(TM, \xi))$ is generic if it intersects each of the subbundles $Hom_r(TM, \xi)$, for $0 \leq r \leq n - 1$, transversely [10] (this is equivalent to saying that its local representatives for one, and hence any, choice of bundle atlases in TM and ξ, regarded as maps $U \to L(n, N)$, intersect each of the submanifolds $L_r(n, N)$ transversely). The term "generic" is justified by the transversality theorem: the space of ξ-valued one-forms which are transverse to $Hom_r(TM, \xi)$ is dense in the space of all ξ-valued one-forms for all $0 \leq r \leq n$. So the set of generic ξ-valued one-forms is a finite intersection of open dense subsets of the space of all ξ-valued one-forms, and since the latter is of the second category in the C^∞-topology (it is a Baire space), it is itself dense.

If θ is generic, a standard result on transversal mappings implies that M_n^θ is a dense open subset of M and $M_{\leq r}^\theta$, for $0 \leq r \leq n - 1$, is a closed submanifold of M of codimension equal to $codim(L_{\leq r}(n, N), L(n, N)) = (n - r)(N - r)$. Depending upon n, N and r, $dimM_{\leq r}^\theta = n(1 - N) + r(n + N) - r^2$ could turn out to be formally negative, in which case $M_{\leq r}^\theta$ is empty. For $r = n - 1$ we see that a transverse intersection is necessarily empty if $N > 2n - 1$, and occurs on a $(2n - N - 1)$-dimensional submanifold if $N \leq 2n - 1$. This is in agreement with the results from obstruction theory. So we get the following picture of a generic θ: if $N > 2n-1$, θ has constant maximal rank n; if $N \leq 2n-1$, θ has constant maximal rank everywhere except for a $(2n-N-1)$-dimensional submanifold $M^{\theta}_{\leq n-1}$ where its rank is $\leq n-1$; if $N \leq \frac{1}{2}(3n-4)$, $M^{\theta}_{\leq n-1}$ contains a $(3n-2N-4)$-dimensional submanifold $M^{\theta}_{\leq n-2}$ where $rank\theta \leq n-2$, a.s.o. The sequence of submanifolds ends when θ reaches a minimal rank, which is the smallest integer $\leq \frac{1}{2}[n+N-\sqrt{(N-n)^2+4n}]$.

A similar discussion could be repeated for κ and g. If κ is generic, then, with obvious notation, M^{κ}_{N} is open and dense in M and $M^{\kappa}_{\leq q}$ for $0 \leq q \leq N-1$ is a closed submanifold of M of dimension $n-(N-q)^2$. Similarly if g is generic, M^{g}_{n} is open and dense in M and $M^{g}_{\leq q_1}$ for $0 \leq q_1 \leq n-1$ is a closed submanifold of M of dimension $n-(n-q_1)^2$.

What can be said about the GLCC's relative to some generic θ and κ? It is clear that there are necessary and sufficient conditions for the existence of a GLCC, similar to those given in the theorem, except that now the dimension of the distributions $ker\theta$ and $kerg$ will be larger on the submanifolds $M^{\theta}_{\leq r}$ and $M^{g}_{\leq q_1}$. Furthermore, the forms α, β, γ and δ which classify the GLCC's will have to satisfy certain boundary conditions on $M^{\theta}_{\leq r}$, $M^{\kappa}_{\leq q}$ and $M^{g}_{\leq q_1}$. The description of the general case, with arbitrary n, N, r, q and q_1, will be rather complicated, and we shall not pursue this any further.

5. Gauge equivalences

The group $\mathcal{A}ut\xi$ of C^∞ linear automorphisms of ξ acts from the right on the geometric objects related to ξ. Let $u \in \mathcal{A}ut\xi$ cover $f \in \mathcal{D}iffM$.

If v is a section of ξ,

$$v \mapsto v' = u^{-1} \circ v \circ f;$$

if X is a vectorfield on M

$$X \mapsto X' = Tf^{-1} \circ X \circ f;$$

if θ is a ξ-valued one-form

$$\theta \mapsto \theta' = u^{-1} \circ \theta \circ Tf;$$

if κ is a bilinear form on ξ,

$$\kappa \mapsto \kappa' = u^* \kappa;$$

if ∇ is a linear connection on ξ

$$\nabla_X v \mapsto \nabla'_X v = \nabla'_{Tf(X)}(u \circ v \circ f^{-1}).$$

The triples (θ,κ,∇) and $(\theta',\kappa',\nabla')$ describe the same physical situation: they are gauge equivalent. We would like to describe the space of gauge inequivalent GLCC's. Since the GLCC's are defined for fixed θ and κ, we are led to ask: are there elements of $\mathcal{A}ut\xi$ which preserve θ and κ and map solutions of the system 1.3, 1.4 to other solutions? From the definitions 2.1, 2.2 we easily get that

$$d_{\nabla'}\theta'(X',Y') = u^{-1}(d_{\nabla}\theta(X,Y)) \qquad 5.1$$

$$\nabla'\kappa' = u^{*}\nabla\kappa \qquad 5.2$$

and so if ∇ is a GLCC with respect to θ and κ, ∇' is a GLCC with respect to θ' and κ'. There follows that given θ and κ, the space of all gauge inequivalent GLCC's is the quotient of the space of all GLCC's, described in part b) of the theorem, by the subgroup $\mathcal{G}(\theta,\kappa) \subset \mathcal{A}ut\xi$ which stabilizes θ and κ. In the rest of this section we describe the group $\mathcal{G}(\theta,\kappa)$. We begin by studying the subgroup $\mathcal{G}(\theta)$ which stabilizes θ alone.

Lemma 1. $u \in \mathcal{A}ut\xi$ **covering** $f \in \mathcal{D}iffM$ **preserves** θ **if and only if** $Tf|_{ker\theta} \in \mathcal{A}ut(ker\theta)$ **and** $u|_{im\theta} = \theta \circ Tf|_{\eta} \circ (\theta|_{\eta})^{-1}$.

u preserves θ if and only if $\theta = u^{-1} \circ \theta \circ Tf$. Decomposing this equation we get $0 = u^{-1} \circ \theta \circ Tf|_{ker\theta}$ and $\theta|_{\eta} = u^{-1} \circ \theta \circ Tf|_{\eta}$. The first implies $Tf(ker\theta) \subset ker\theta$, and since $Tf \in \mathcal{A}utTM$, $Tf|_{ker\theta} \in \mathcal{A}ut(ker\theta)$. The second implies that $u^{-1}|_{im\theta} \subset im\theta$ and since $u^{-1} \in \mathcal{A}ut\xi$, $u^{-1}|_{im\theta} \in \mathcal{A}ut(im\theta)$. But then $u^{-1}|_{im\theta} = (u|_{im\theta})^{-1}$ and so composing on the left with $u|_{im\theta}$ and on the right with $\theta|_{\eta}$ we get $u|_{im\theta} = \theta \circ Tf|_{\eta} \circ (\theta|_{\eta})^{-1}$. This proves the lemma.

Let $\mathcal{H}om_f(\alpha,\beta)$ denote the set of vectorbundle homomorphisms from α to β covering $f \in \mathcal{D}iffM$ (not to be confused with the vectorbundle $Hom(\alpha,\beta)$, whose sections are $\mathcal{H}om_{Id_M}(\alpha,\beta)$). Let $\mathcal{H}om(\alpha,\beta) = \cup_{f\in\mathcal{D}iffM}\mathcal{H}om_f(\alpha,\beta)$.

Corollary 1. $\mathcal{G}(\theta) = \cup_f(\mathcal{A}ut_f\varsigma \times \mathcal{H}om_f(\varsigma, im\theta))$, **the union running over all diffeomorphisms which preserve the distribution defined by** $ker\theta$.

This follows from the observation that $u|_{im\theta} \in \mathcal{A}ut(im\theta) \subset \mathcal{H}om(im\theta,\xi)$ is determined by f and $u|_{\varsigma}$ is free, except for the requirement that $\Pi_{\varsigma} \circ u|_{\varsigma} \in \mathcal{A}ut\varsigma$.

We now turn to the group $\mathcal{G}(\kappa)$ which stabilizes κ.

Lemma 2. $u \in \mathcal{A}ut\xi$ **preserves** κ **if and only if** $u|_{\xi_0} \in \mathcal{A}ut\xi_0$ **and** $\Pi_{\xi_+} \circ u|_{\xi_+} \in \mathcal{A}ut^{O(q)}\xi_+$.

Here $\mathcal{A}ut^{O(q)}\xi_+$ denotes the subgroup of $\mathcal{A}ut\xi_+$ which preserves $\kappa|_{\xi_+}$. The elements of $\mathcal{G}(\kappa)$ are represented locally by maps from U to the group of matrices of the form 4.1.

Corollary 2. $\mathcal{G}(\kappa) = \cup_f(\mathcal{A}ut_f^{O(q)}\xi_+ \times \mathcal{H}om_f(\xi_+, \xi_0) \times \mathcal{A}ut_f\xi_0)$, **the union running over all** $f \in \mathcal{D}iff M$.

Finally, we can tackle the combined problem.

Lemma 3. $u \in \mathcal{A}ut\xi$ **covering** $f \in \mathcal{D}iff M$ **preserves** θ **and** κ **if and only if** $u|_{im\theta_+} \in \mathcal{H}om(im\theta_+, im\theta)$; $\Pi_{im\theta_+} \circ u|_{im\theta_+} \in \mathcal{A}ut^{O(q_1)}(im\theta_+)$; $u|_{im\theta_0} \in \mathcal{A}ut(im\theta_0)$; $\Pi_{\varsigma_+} \circ u|_{\varsigma_+} \in \mathcal{A}ut^{O(q_2)}\varsigma_+$; $u|_{\varsigma_0} \in \mathcal{H}om(\varsigma_0, \xi_0)$; $\Pi_{\varsigma_0} \circ u|_{\varsigma_0} \in \mathcal{A}ut\varsigma_0$; f **is an isometry of** g; $Tf|_{ker\theta} \in \mathcal{A}ut(ker\theta)$; $Tf|_{kerg} \in \mathcal{A}ut(kerg)$; $\Pi_{\eta_+} \circ Tf|_{\eta_+} \in \mathcal{A}ut^{O(q_1)}\eta_+$; $u|_{im\theta} = \theta \circ (Tf|_\eta) \circ (\theta|_\eta)^{-1}$.

Proof. The first two statements follow directly from lemma 1; the third statement follows from lemmas 1 and 2 together; $u|_{\varsigma_+} \in \mathcal{H}om(\varsigma_+, \xi)$ with the only restriction that its projection onto ς_+ is orthogonal; the fifth and sixth conditions follow from lemma 2. From lemma 1, $u \circ \theta = \theta \circ Tf$, so $Tf^*g = Tf^*\theta^*\kappa = \theta^*u^*\kappa = \theta^*\kappa = g$; furthermore $Tf|_{ker\theta} \in \mathcal{A}ut(ker\theta)$ and $u|_{im\theta} = \theta \circ (Tf|_\eta) \circ (\theta|_\eta)^{-1}$. To prove the remaining two statements, we observe that $\theta|_\eta : \eta \to im\theta$ is an isomorphism and $(u|_{im\theta}) \circ (\theta|_\eta) = (\theta|_\eta) \circ (Tf|_\eta)$; it follows from the first three conditions on u that $Tf|_{\eta_+} \in \mathcal{H}om(\eta_+, \eta)$; $\Pi_{\eta_+} \circ Tf|_{\eta_+} \in \mathcal{A}ut^{O(q_1)}\eta_+$; $Tf|_{\eta_0} \in \mathcal{H}om(\eta_0, \eta_0 \oplus ker\theta)$. Because of 2.13, $Tf|_{kerg} \in \mathcal{A}ut(kerg)$.

Corollary 3. $\mathcal{G}(\theta, \kappa) = \cup_f[\mathcal{A}ut_f^{O(q_1)}(im\theta_+) \times \mathcal{H}om_f(im\theta_+, im\theta_0) \times \mathcal{A}ut_f(im\theta_0) \times \mathcal{A}ut_f^{O(q_2)}\varsigma_+ \times \mathcal{H}om_f(\varsigma_+, im\theta \oplus \varsigma_0) \times \mathcal{A}ut_f\varsigma_0 \times \mathcal{H}om_f(\varsigma_0, im\theta_0)]$, **the union running over all isometries of** g **which preserve the distributions defined by** $ker\theta$ **and** $kerg$.

References

[1] R. Percacci, "Geometry of nonlinear field theories", World Scientific(1986).
[2] R. Percacci ,"Role of soldering in gravity theory", in the proceedings of the XIII DGM Conference, Eds. H.D. Doebner and T.D. Palev, World Scientific (1986).
[3] A. Einstein, W. Meyer, *Sitzungsber. der Preuss. Akad. der Wiss.* (1931) p.141; *ibid.* (1932) p.130.
[4] N. Rosen and G. Tauber, *Found of Phys.***14** (1984) 171.
[5] R. Percacci, *Phys. Lett.***144B** (1984) 37.
[6] G. Jankiewicz, *Bull.Acad.Polon.Sci.*, vol.**II** (1954) 301.
[7] W.O. Vogel, *Arch. Mat.*, vol.**XVI** (1965) 106.
[8] M. Crampin, *Proc. Camb. Phil. Soc.***64** (1968) 64.
[9] N. Steenrod, "The topology of fibre bundles" Princeton Univ. Press (1951).
[10] M. Golubitski and V. Guillemin, "Stable mappings and their singularities", Springer (1973).

SUR LA FORMULATION GEOMETRIQUE DES THEORIES DE JAUGE

J. SHABANI

Département de Mathématique, Université du Burundi,

Bujumbura, Burundi.

RESUME: Nous passons en revue quelques aspects fondamentaux de la géometrie des théories de jauge. En particulier, nous introduisons les concepts de potentiel de jauge, intensité de champ, champ de matière, groupe de jauge et symétrie d'une configuration physique et discutons le mecanisme de brisure spontanée de symétrie et les théories de jauge de la gravitation.

1. Introduction

L'aspect géométrique des théories de jauge, basé sur la théorie des connexions et des courbures sur un fibré principal, a joué un rôle important dans la description et l'étude des phénomènes comme : la symétrie des configurations physiques [1], les monopoles magnétiques [2], les spineurs [3], les anomalies [4], les nombres quantiques topologiques, les instantons [5], etc ....

Dans cet exposé, nous passons en revue certains aspects importants de la géométrie des théories de jauge. Nous nous limitons aux théories classiques, car ce sont celles-là qui se prêtent le mieux à une interprétation géométrique. Pour les théories de jauge quantiques, developpées notamment par Faddeev-Popov [6] et De Witt [7] nous réferons le lecteur au livre de Konopleva et Popov [5].

Toutes les notions de géometrie différentielle que nous utilisons sont discutees en détail dans [8,9]

Dans la Section 2, nous donnons les différentes définitions de base de la géométrie d'une théorie de jauge. La Section 3 est consacrée à l'étude du groupe d'automorphismes d'un fibré principal P (groupe de jauge de la théorie de jauge associée à P) et nous discutons la notion de symetrie d'une configuration physique.

Dans la section 4, nous étudions le mécanisme de brisure spontanée de symétrie et quelques unes de ses conséquences. La Section 5 passe en revue les différentes théories proposées comme théories de jauge de la gravitation et enfin dans la Section 6 nous donnons une très brève introduction aux théories de jauge supersymétriques.

2. Définitions fondamentales

Dans ce paragraphe nous passons en revue les fondements des théories de jauge dans le langage des espace fibrés. Nous suivrons dans la mesure du possible la terminologie et les notations de Kobayashi-Nomizu [9] (voir aussi [10]).

Soit $P \equiv (P,M,G,\pi)$ un fibré principal sur (l'espace-temps) M. Le groupe de structure G est parfois appelé groupe de jauge. Suivant [1], nous appelerons groupe de jauge d'une théorie, le groupe d'automorphismes de P laissant invariants **les éléments** absolus de cette **théorie** (voir section 3).
Soit ω une 1-forme de connexion sur P et $s: U \in M \to P$ une section locale de P. La section s et la 1-forme ω déterminent une 1-forme A sur M à valeurs dans l'algèbre de Lie $\mathfrak{g}$ de G appelée *potentiel de jauge* et définie par le pull-back de ω par s, c'est-à-dire :

$$A = s^*\omega \tag{2.1}$$

On appelle *intensité de champ* la 2-forme à valeurs dans $\mathfrak{g}$ définie par

$$F = s^*\Omega \tag{2.2}$$

où Ω est la 2-forme de courbure de la 1-forme ω.
Notons en passant que Ω vérifie l'*identité de Bianchi :*

$$D\Omega \equiv d\Omega + [\omega,\Omega] = 0 \tag{2.3}$$

D'habitude les physiciens travaillent avec des potentiels et des champs exprimés dans une jauge locale et ceci justifie le fait de considérer une section locale dans les définitions de A et F. Notons que la formulation présentée ci-dessus peut être étendue à tout l'espace-temps.

Une autre notion fondamentale en théorie de jauge est celle de *champ de matière*. Cette notion peut être introduite de deux façons différentes.
Soit V un espace vectoriel sur lequel G agit par la gauche par une représentation linéaire $\rho: G \to GL(V)$. On définit le champ de matière associé à cette représentation comme étant une fonction $\phi: P \to V$ équivariante sous l'action

de G, c'est-à-dire vérifiant la relation

$$\phi(y.g) = \rho(g^{-1})\phi(y) \quad ; \quad \forall y \in P, g \in G \tag{2.4}$$

De même on peut définir le champ de matière comme étant une section du fibré vectoriel associé à P, avec fibre standard V. Ce fibré vectoriel noté $E \equiv E(M,V,G,P)$ est défini comme suit :

Le groupe G agit de façon naturelle par la droite sur le produit cartésien $P \times V$; cette action étant donnée par :

$$(y,v).g = (y.g, \rho(g^{-1})v) \quad ; \quad \forall y \in P \ , v \in V, \ g \in G. \tag{2.5}$$

L'orbite de cette action de G sur $P \times V$, notée $P \times_G V$ est le fibré associé à P par la représentation ρ, avec fibre standard V.

De la même façon, on peut définir un fibré vectoriel associé à P, avec fibre standard F (où F est une variété arbitraire) sur lequel G agit par la gauche. En particulier le fibré associé $E \equiv E(M, G/H, G, P)$ est utilisé dans l'étude de certains modèles rencontrés en physique. Une section d'un tel fibré est identifiée à un champ à valeurs dans l'espace homogène G/H .

Notons que le cas particulier $G = SO(3)$, $H = SO(2)$ c-à-d $G/H = S^2$, correspond aux modèles σ classiques. Plus généralement les théories de champ basées sur les sections d'un fibré vectoriel avec fibre standard un espace homogène sont appelées modèles sigma généralisés. Un exemple d'une telle situation est fourni par les modèles CP^{n-1}.

Si le champ de matière tel que défini ci-dessus est un champ scalaire, on l'appelle *champ de Higgs* de type (ρ,V). Un champ de Higgs est dit *standard* s'il est du type $(Ad,\mathfrak{g})$ où $\mathfrak{g}$ est l'algèbre de Lie de G.

Une des questions fondamentales est celle de savoir quand est-ce qu'une théorie de jauge est définie.

Soit $P = (P,M,G,\pi)$ un fibré principal sur M. Une théorie de jauge est définie si l'on spécifie au moins les éléments suivants :

(i) L'espace-temps M (muni d'une structure de variété différentiable) c-à-d. l'arène dans laquelle se tiendront les phénomènes que la théorie veut décrire.

(ii) Le groupe de structure G du fibré P.

(iii) Le type de particules couplés au champ de jauge. Le type de particules en considération est déterminé par le choix d'une représentation $\rho: G \to GL(V)$

(iv) La forme des équations de champ. Ces équations peuvent s'obtenir d'une densité de Lagrangien (une 4-forme) en utilisant le principe variationnel habituel.

Ce schéma nous sera très utile à la Section 5 pour la classification de différentes structures proposées dans l'étude de la gravitation comme théorie de jauge.

3. Groupes de jauge et symétries des configurations physiques

Soit $P = (P,M,G,\pi)$ un fibré principal sur M. Dans ce paragraphe on introduit la notion de transformation de jauge et celle de symétrie qui est définie comme étant une transformation de jauge u qui préserve la 1-forme de connexion ω. On montre que l'invariance de ω sous l'action de u est équivalente à l'annulation de la dérivée de Lie de ω par rapport au champ de vecteur sur P correspondant à un groupe à un paramètre d'automorphismes de P.

Un difféomorphisme $u: P \to P$ est un *automorphisme* de P s'il existe un difféomorphisme $v: M \to M$ tel que $\pi.u = v.\pi$ et $u(pa) = u(p)a$; $\forall p \in P$, $a \in G$. On dénote par Aut P le groupe de tous les automorphismes de P.
Le difféorphisme v est déterminé univoquement par l'automorphisme u et on a l'homomorphisme de groupe :

$$\begin{aligned} j \ : \ \mathrm{Aut}\ P &\longrightarrow \mathrm{Diff}\ M \\ u &\longmapsto j(u) = v \end{aligned} \tag{3.1}$$

Un automorphisme est dit *vertical* si $j(u) = id_M$. L'ensemble $\mathrm{Aut}_M\ P$ des automorphismes verticaux est un sous-groupe normal de Aut P et on a la suite exacte :

$$1 \longrightarrow \mathrm{Aut}_M\ P \xrightarrow{\ i\ } \mathrm{Aut}\ P \xrightarrow{\ j\ } \mathrm{Diff}\ M = \mathrm{Aut}\ P /_{\mathrm{Aut}_M\ P} \longrightarrow 1 \tag{3.2}$$

où i est l'injection canonique.

Dans une théorie de jauge, il est parfois nécessaire de distinguer les variables dynamiques comme ω ou le champ de Higgs, des éléments absolus comme le tenseur métrique en relativité spéciale [11].
Le *groupe de jauge* d'une telle théorie est par définition le sous-groupe $\mathcal{G}$ de Aut P laissant invariants les éléments absolus.
Les éléments de $\mathcal{G}$ sont appelés *transformations de jauge*, et les éléments verticaux de $\mathcal{G}$ sont les *transformations de jauge pures*. Le *groupe de jauge pur* $\mathcal{G}_o = \mathcal{G} \cap \mathrm{Aut}_M P$ est un sous-groupe normal de $\mathcal{G}$ et on a la suite exacte :

$$1 \longrightarrow \mathcal{G}_o \xrightarrow{i} \mathcal{G} \xrightarrow{j} \mathcal{G}/\mathcal{G}_o \longrightarrow 1 \tag{3.3}$$

où $\mathcal{G}/\mathcal{G}_o$ est identifié à $j(\mathcal{G}) \subset \mathrm{Diff}\ M$

Si $u \in \mathrm{Aut}_M P$, alors p et u(p) appartiennent à la même fibre. Dès lors, il existe un élément $U(p) \in G$ tel que $u(p) = pU(p)$ et :

$$U(pa) = a^{-1}U(p)a \quad ; \quad \forall p \in P \ , \quad a \in G \tag{3.4}$$

Ainsi, il existe un isomorphisme naturel de $\mathrm{Aut}_M P$ sur le groupe des applications $U : P \to G$ vérifiant la condition d'équivariance (3.4).

Soit P un fibré principal, ω-une 1-forme de connexion sur P et $u \in \mathrm{Aut}\ P$. L'invariance de ω sous l'action de u signifie que le pull-back de ω par u, c-à-d. $\omega' = u^*\omega$ est de nouveau une connexion sur P.
Si $u \in \mathrm{Aut}_M P$, alors :

$$\omega' = \mathrm{Ad}(U^{-1}(p))\omega + U^{-1}(p)\ dU(p) \tag{3.5}$$

et
$$\Omega' = u^*\omega' = \mathrm{Ad}(U^{-1}(p))\Omega \tag{3.6}$$

Soit s une section locale de P. Nous allons montrer qu'un changement de section correspond à une transformation de jauge habituelle pour $A = s^*\omega$ et $F = s^*\Omega$. En effet, le potentiel transformé $A' = s'^*\omega = s^*\omega'$ peut être interprété comme étant le pull back de ω par $s' = u.s$ ou le pull-back de $\omega' = u^*\omega$ par s.

En posant $S = U.s$, on obtient de (3.5) et (3.6) les formules usuelles de transformation de jauge :

$$A' = Ad(S^{-1})A + S^{-1}dS \ , \quad F' = Ad(S^{-1}F) \tag{3.7}$$

On appelle *symétrie* d'une configuration physique, une transformation de jauge u qui préserve ω, c-à-d. telle que :

$$u^*\omega = \omega \tag{3.8}$$

Nous allons montrer que la condition (3.8) peut s'exprimer en termes de l'annulation de la dérivée de Lie de ω.

Considérons un groupe à un paramètre (u_t), $t \in R$ d'automorphismes de P. Soit Z le champ de vecteurs sur P correspondant à (u_t) et ξ la projection de Z sur M. Le champ de vecteurs ξ engendre un groupe de transformations à un paramètre (v_t), $v_t = j(u_t)$ sur M.

La 1-forme de connexion ω est préservée par u_t si et seulement si la dérivée de Lie par rapport à Z s'annule, c-à-d. $L_Z\omega = 0$.

4. Brisure spontanée de symétrie (BSS) et classification des théories de jauge

La plupart des théories de jauge intéressant les physiciens comme l'unification électrofaible [12] caractérisée par le groupe de structure $SU(2) \times U(1)$ ou les différentes théories de grande unification, par exemple celles basées sur $SU(5)$, $SO(10)$ et sur les différents groupes de Lie exceptionnels font appel au mécanisme de BSS proposé par Higgs en 1964 [13] .

Dans ce paragraphe, nous décrivons le mécanisme de BSS et nous montrons comment ce mécanisme permet de construire une hiérarchie des théories de jauge définies sur le même espace-temps. La comparaison des groupes de structure associés à ces différentes théories permet alors de les classifier. Il faut quand même noter que cette classification doit tenir compte du lien existant entre les symétries des solutions des équations de champ (= les 1-formes de connexion sur le fibré principal attaché à la théorie) et les groupes d'holonomies de ces connexions.

Soit $P = (P,M,G,\pi)$ un fibré principal sur M. Considérons un champ de Higgs à valeurs dans une orbite W de G dans l'espace vectoriel V des représentations de G, c-à-d. :

$$\alpha:\ P \to W \subset V \tag{4.1}$$

et W est tel que $\forall w, w_\circ \in W$, $\exists a \in G$ tel que

$$w = \rho(a)w_\circ \tag{4.2}$$

Soit $H = \{a \in G \mid \rho(a)w_\circ = w_\circ\}$ le groupe d'isotropie de $w_\circ$. Alors [9] :
$Q = \{p \in P \mid \alpha(p) = w_\circ\}$ est un sous-espace fibré de P sur M. De plus $Q \equiv (Q,M,H,\pi)$ possède une structure de fibré principal avec groupe de structure H.
Dans une théorie de jauge associé au fibré principal P, le passage de P à Q est appelé *réduction* du fibré P à Q ou *brisure spontanée de symétrie*.

Un exemple de BSS peut être obtenu en considérant la solution de 't Hooft-Polyakov [14] des équations de Yang-Mills couplées à un champ de Higgs de type Ad. Dans ce cas $P = M\times G = S^2 \times SO(3)$ et le champ de Higgs $\alpha: S^2 \times SO(3) \to S^2$ est donné par :

$$\alpha(r,a) = a^{-1}r \quad ; \quad r = (x,y,z) \in S^2 , \quad a \in SO(3) \tag{4.3}$$

Soit $r_\circ = (0,0,1)$. Alors $H = \{a \in SO(3) \mid ar_\circ = r_\circ\} = SO(2)$,
$Q = \{(r,a) \in P \mid a^{-1}r = r_\circ\}$ et le champ α réduit le fibré trivial P au fibré non trivial Q de groupe de structure H.

La théorie des réductions et extensions des fibrés principaux permet de construire une hiérarchie de théories de jauge à partir d'un principe très simple: changer le groupe de structure pour passer d'une théorie à l'autre. En effet, donnée une interaction caractérisée par son groupe de structure, le passage à une interaction forte correspond à une extension du groupe de structure, tandis que le passage à une interaction faible conduit à une brisure de symétrie et donc à une réduction du groupe de structure.
Cependant la réalisation de cette classification des théories de jauge doit tenir compte du fait qu'il existe une dépendance entre l'espace-temps et les symétries internes des solutions des équations de champ (c-à-d les symétries des connexions

sur le fibré principal attaché à la théorie considérée) via le *groupe d'holonomie* qui est un sous-groupe du groupe de structure.

Ainsi, dans une théorie de jauge pure (c-à-d sans source), demander que la densité d'énergie du champ soit définie positive revient à exiger que les groupes d'holonomie des solutions des équations de champ soient compacts et semi-simples [5] .

5. Gravitation comme théorie de jauge

Malgré qu'il existe quelques différences entre la gravitation et les théories de jauge classiques comme l'électromagnétisme et la théorie de Yang-Mills basée sur SU(2) (pour une comparaison entre ces théories, nous réferons le lecteur à[15]), plusieurs études ont été faites (voir par exemple [16]) pour essayer d'incorporer la gravitation dans le schéma général des théories de jauge. Le point de départ de ces études est le fibré principal $P = (P,M,G,\pi)$ sur une variété pseudo-Riemanienne M. Les différentes théories proposées sont basées sur différents choix du groupe de structure G.

Dans ce paragraphe, nous passons en revue ces différentes théories. Pour plus de détails nous réferons à [16,17] .

5.1. *G est le groupe des translations* T(4)

Dans ce cas P est le fibré trivial $P = M\times T(4)$. Une telle théorie a été étudiée par Cho [18] (et est parfois appelée théorie de Cho) et par Hayashi et Nakano [19] .

5.2. *G est le groupe de Lorentz* SO(3,1)

Ici P est le fibré principal des repères orthogonaux sur M et la théorie obtenue contient comme cas particuliers la théorie d'Einstein avec terme cosmologique et la théorie du type Einstein-Cartan avec terme cosmologique [20] .

5.3. *G est le groupe* GL(4, R)

Dans ce cas P est le fibré principal des répères généraux sur M. Une telle théorie a été étudiée par Yang [21] et contient comme cas particuliers la théorie d'Einstein avec terme cosmologique et les théories de type Einstein-Cartan avec terme cosmologique, Yang-Stephenson [21,22] et Mansouri-Chang [23] .

5.4. *G est le groupe de Poincaré*

Dans ce cas, P est le fibré principal des repères affines orthogonaux sur M.

5.5. *G est le groupe de De Sitter*

Ici P est le fibré principal des repères affines et la théorie obtenue est appelée théorie de Mac Dowell-Mansouri [24].

Suivant la procédure décrite ci-dessus, d'autres théories ont été proposées, notamment celles basées sur les groupes produit direct $T(4)\times SO(3,1)$ et $T(4)\times GL(4,R)$. Pour plus de détails sur ces théories ainsi que leur interprétation en termes de la géométrie des espaces fibrés, nous réferons le lecteur à [16] .

6. Théories de jauge supersymétriques

La brisure spontanée de symétrie introduit un paramètre arbitraire λ (constante de couplage des champs scalaires). Ce paramètre peut être fixé si nous demandons que la théorie soit supersymétrique.
Pour l'étude de telles théories, Kostant [25] a introduit la notion de variété graduée. Cette notion qui a été à la fois raffinée et simplifiée par Koszul peut être utilisée pour la construction d'une supervariété [26], notion qui joue en supergravité le rôle joué par une variété différentiable en relativité générale.

Nous n'allons nous étendre sur ce domaine qui fait pour l'instant l'objet de recherches intensives de la part de Mathématiciens et Physiciens théoriciens.

Références

[1] A.TRAUTMAN : Bull. Acad. Polonaise Sci, Série Sci Phys et Astron , Vol 27 n°1,7 (1979).

[2] P.A.HORVATHY : J. Geom. and Phys., vol 1 , n°3, 39 (1984)

[3] L.DABROWSKI et R.PERCACCI : Comm. Math. Phys. 106 , 691 (1986)

[4] R.PERCACCI : voir les présents Proceedings.

[5] N.P.KONOPLEVA et V.N.POPOV : Gauge Fields , Harwood Academic Publishers , Chur - London - New York 1981.

[6] L.D.FADDEEV et V.N.POPOV : Phys.Lett.25 B , 29 (1967)

[7] B.S. De WITT : Phys.Rev.162 , 1195 , 1239 (1967)

[8] A. LICHNEROWICZ : Théorie globale des connexions et des groupes d'holonomie, Edizioni Cremonese, Roma, 1962.

[9] S.KOBAYASHI et K.NOMIZU : Foundations of Differential Geometry, Interscience, New York , vol I : 1963 ; vol II : 1969.

[10] A.TRAUTMAN : Rep. Math. Phys.1 , 29 (1970) ; W.DRECHSLER et M.E.MAYER : Fiber Bundle Techniques in Gauge Theories. Lecture Notes in Physics 67, Springer-Verlag , Berlin-Heidelberg-New York 1977 ; M.E.MAYER et A.TRAUTMAN : Acta Phys.Austr., Suppl 23 , 433-476 (1981).

[11] J.L.ANDERSON : Principles of Relativity Physics, Academic Press, New York 1967 , §4.3.

[12] S.WEINBERG : Phys.Rev. Lett.19 , 1264 (1967) ; A.SALAM in Proc 8th Nobel Symposium , Stockholm (1968) p.367.

[13] P.W. HIGGS : Phys.Lett.12 , 132 (1964).

[14] G.'t HOOFT : Nucl.Phys.B 79, 276 (1974); A.POLYAKOV: Zh Eksper. Teor. Fiz. Pis'ma Red 20, 430 (1974).

[15] A.TRAUTMAN : Yang-Mills theory and gravitation : a comparison. Lecture Notes in Mathematics n°926, pp 179-189. Springer, Berlin 1982; W.THIRRING: Acta Phys.Austr., Suppl 19, 439 (1978).

[16] F.G.BASOMBRIO : Gen.Rel.and Gravitation 12 , 109 (1980)

[17] D.IVANENKO et G.SARDANASHVILY: Phys.Rep.94, pp 1-45 (1983); T.KAWAI: Gen. Rel.and Gravitation 18, 995 (1986).

[18] Y.M.CHO : Phys.Rev.D 14, 2521 (1976)

[19] K.HAYASHI et T.NAKANO : Progr.Theor.Phys.38, 491 (1967)

[20] A.TRAUTMAN : "The Einstein-Cartan Theory" in Proc 9th Int. Conf on General Relativity and Gravitation. Jena 1980, pp 225-227. Edited by E.SCHMUTZER, Cambridge University Press, Cambridge 1983.

[21] C.N.YANG : Phys.Rev.Lett.33, 445 (1974)

[22] G.STEPHENSON: Nuovo Cim.9 , 263 (1958)

[23] F.MANSOURI et L.N.CHANG : Phys.Rev. D 13 , 3192 (1976)

[24] S.MAC DOWELL et F.MANSOURI : Phys.Rev.Lett.38, 739 (1977).

[26] B.KOSTANT : "Graded manifolds, graded Lie groups and prequantisation". Lecture Notes in Mathematics, vol.570 pp 177-306, Springer, Berlin-Heidelberg-New York (1977)

[26] A.ROGERS : J. Math.Phys.21, 1352 (1980) ; Comm.Math.Phys.105, 375 (1986).

III: Sur les opérateurs differentiels et pseudo-differentiels et analyse

OPERATEURS DIFFERENTIELS NATURELS SUR LES FIBRES VECTORIELS

Jean Pierre EZIN

Département de Mathématiques,
Université Nationale du Bénin, Cotonou, Bénin.

Abstract. This paper is an expository talk which presents concepts involved in proving the following proposition: on a closed Riemannian 4-manifold M equipped with a conformal class γ with positive Yamabe number μ_γ, every Yang-Mills field R^∇ on a SU(2)-bundle over M is self-dual whenever $48\int_M |R_-^\nabla|^2 < \mu_\gamma$.

Nous présentons dans cet exposé les considérations élémentaires conduisant à un théorème d'isolation L^2 (Th.3.7) pour des champs de Yang-Mills i.e. des tenseurs de courbure R^∇ de G-connexions critiques ∇ de la fonctionnelle de Yang-Mills, où G est un sous-groupe compact du groupe linéaire réel. Ce résultat est extrait de [6] où plus d'informations sur les motivations du théorème peuvent être obtenues.

Le plan de l'exposé est le suivant. La Section 1, à caractère didactique, donne la décomposition classique de l'opérateur de courbure d'un espace tangent V à une variété riemannienne (M,g) en composantes O(V)-irréductibles. Elle peut être considérée comme une section de référence et n'a pas besoin d'être systématiquement lue. Dans la Section 2 nous écrivons une formule de Weitzenböck comparant deux laplaciens naturels opérant sur les 2-formes à valeurs vectorielles sur (M,g) telle qu'elle figure dans [5] dont nous utilisons librement définitions et résultats. C'est sur cette formule que repose la preuve du théorème d'isolation, objet de la Section 3 où nous introduisons le nombre de Yamabe de la classe conforme γ de la métrique g.

1. PRELIMINAIRES. NOTATIONS

1.1. Soit $(V,<,>)$ un espace vectoriel euclidien de dimension $n \geqslant 2$. L'espace vectoriel $\Lambda^k(V)$ (ou Λ^k) de ses k-vecteurs est muni d'un produit scalaire induit <u>noté</u> encore $<,>$ et donné sur les éléments décomposables par $<u_1 \wedge \ldots \wedge u_k, v_1 \wedge \ldots \wedge v_k> = \text{dét}\,[<u_i,v_j>]$ pour u_i, v_j dans V. Si (e_i), $i = 1, \ldots n$, est une base orthonormée de V, la base correspondante de $\Lambda^k(V)$ est $(e_{i_1} \wedge \ldots \wedge e_{i_k})$ $i_1 < \ldots < i_k$ et elle est orthonormée. Le produit scalaire $<,>$ étant fixé une fois pour toutes sur V, on identifiera V et $\Lambda^k(V)$ avec leurs duaux respectifs.

1.2. On munit l'espace vectoriel End $\Lambda^k(V)$ des endomorphismes symétriques de Λ^k d'une structure euclidienne en posant:

$(T,S) = \text{tr}\,(T \circ S)$, pour T,S dans End Λ^k où tr désigne la trace d'une application linéaire de Λ^k dans Λ^k.

Dans le cas particulier où $k = 2$, on <u>note</u> $S^2\Lambda^2(V)$ le sous-espace de dimension $\frac{1}{2} C_n^2 (1 + C_n^2)$ de End Λ^2 constitué par les endomorphismes symétriques. Un élément de $S^2\Lambda^2(V)$ est appelé un <u>opérateur du type de la courbure</u> sur V.

1.3. A cause de sa symétrie et de l'antisymétrie du produit extérieur Λ, chaque opérateur T du type de la courbure peut être identifié à une 2-forme $\hat{T}$ sur V à valeurs dans les endomorphismes antisymétriques de V par

$$<\hat{T}(x,y)u, v> = <T(x \wedge y), u \wedge v>, \text{ pour } x, y, u, v \text{ dans } V.$$

Nous écrirons souvent T_{xy} pour signifier $\hat{T}(x,y)$ et nous utiliserons la convention que pour ω dans $\Lambda^2(V)$ (compris indifféremment comme espace de bi-vecteurs ou espace de 2-formes)

$$\omega o\hat{T} \equiv \omega o T \equiv T(\omega) = \frac{1}{2} \sum_{i,j=1}^{n} \omega_{ij} T_{e_i e_j}$$

où ω_{ij} sont les composantes de ω dans la base orthonormée $(e_i \wedge e_j)$, $1 \leq i < j \leq n$, de $\Lambda^2(V)$.

La trace de Ricci de T est l'endomorphisme symétrique r de V donné par $x \to r(x) = \sum_{i=1}^{n} T_{e_i x}(e_i)$. La trace de r est la courbure scalaire de T.

1.4. Nous appellerons application de Bianchi l'opérateur linéaire $b: S^2\Lambda^2 \to S^2\Lambda^2$ défini par

$$(b(T))_{x\,y}u = T_{xy}u + T_{yu}x + T_{ux}y$$

Nous désignerons par $\mathcal{R}$ le noyau de b. On se réfère à l'identité

$$(1.5) \qquad (b(R))_{xy}(u) = R_{xy}u + R_{yu}x + R_{ux}y \equiv 0$$

qui définit l'appartenance de R à $\mathcal{R}$ comme la première identité de Bianchi

1.6. Un endomorphisme dans $S^2\Lambda^2$ qui vérifie la première identité de Bianchi (1.5) est appelé opérateur de courbure. Notons $\mathcal{E}$ le complément orthogonal de $\mathcal{R}$ dans $S^2\Lambda^2(V)$. Il est facile de voir que l'application $\omega \to S_\omega$ qui à la 4-forme ω sur V associe l'opérateur du type de la courbure S_ω donné par

$$\langle S_\omega(x \wedge y),\, u \wedge v\rangle = \omega(x,y,u,v), \quad x,\, y,\, u,\, v, \text{ dans } V$$

est un isomorphisme de $\Lambda^4(V)$ (identifié à son dual) sur $\mathcal{E}$. Il suit que $\mathcal{E}$ ne contient que des endomorphismes à courbure sectionnelle (voir exposé II pour une définition) nulle et les contient tous. Dans la suite $\mathcal{E}$ sera identifié à $\Lambda^4(V)$ et nous supposerons $n = \dim V \geqslant 4$.

1.7. En géométrie riemannienne, le tenseur de courbure d'une métrique d'une varieté M induit sur l'espace tangent V en un point de M un opérateur de courbure. Dans certains cas les opérateurs de courbure ainsi obtenus déterminent la géométrie de M; par exemple S. Gallot et D. Meyer ont donné dans [11] une classification complète des variétés riemanniennes compactes, simplement connexes à opérateur de courbure défini semi-positif sur chaque espace tangent. On conjecture même qu'une variété riemannienne compacte simplement connexe à opérateur de courbure positif sur chaque espace tangent est difféomorphe à une sphère standard.

1.8. Pour obtenir la décomposition irréductible sous l'action du groupe orthogonal $O(V)$ du noyau $\mathcal{R}$ de b, il est courant maintenant d'utiliser le formalisme de l'algèbre de courbure [14] [6]. On munit canoniquement le carré tensoriel $\Lambda(V) \otimes \Lambda(V)$ de l'algèbre extérieure $\Lambda(V)$ de V d'une structure d'algèbre en définissant une multiplication $\bullet$ par:

$$(x_1 \otimes y_1) \bullet (x_2 \otimes y_2) = (x_1 \wedge x_2) \otimes (y_1 \wedge y_2).$$

La sous-algèbre

$$KV = \mathbb{R} \oplus S^2V \oplus S^2\Lambda^2(V) \oplus \dots \oplus S^2\Lambda^n(V)$$

de $\Lambda(V) \otimes \Lambda(V)$ est l'<u>algèbre de courbure</u> de V. C'est une algèbre commutative graduée. Le produit $\cdot$ restreint à KV sera noté $\bigcirc\!\!\!\!\wedge$. Ainsi pour h et h' deux formes bilinéaires symétriques sur V, $h \bigcirc\!\!\!\!\wedge h'$ appartient à $S^2\Lambda^2(V)$ et

$$(h \bigcirc\!\!\!\!\wedge h')(x,y) = h(x,\cdot) \wedge h'(y,\cdot) + h(y,\cdot) \wedge h'(x,\cdot), \quad \text{pour } x,y \text{ dans } V.$$

En particulier si g désigne encore le produit scalaire induit sur un espace tangent V à une variété riemannienne (M,g), on a $g \bigcirc\!\!\!\!\wedge g = 2\mathrm{Id}_{\Lambda^2(V)}$.

1.9. Il en résulte la décomposition orthogonale immédiate de $\mathcal{R}$

$$\mathcal{R} = S_0^2\Lambda^2(V) \oplus \mathbb{R}\, g \bigcirc\!\!\!\!\wedge g$$

où $S_0^2\Lambda^2(V)$ est le sous-espace de $\mathcal{R}$ formé des endomorphismes symétriques de $\Lambda^2(V)$ à trace nulle.

Sous l'action du groupe orthogonal $O(V)$, $S_0^2\Lambda^2(V)$ se décompose en deux sous-espaces orthogonaux notés $\mathcal{Z}$ et $\mathcal{W}$. Le noyau $\mathcal{R}$ s'ecrit alors $\mathcal{R} = \mathcal{Z} \oplus \mathcal{W} \oplus \mathcal{U}$, où nous avons écrit $\mathcal{U}$ pour $\mathbb{R} g \bigcirc\!\!\!\!\wedge g$.

Cette décomposition est $O(V)$-irréductible. Les éléments dans $\mathcal{W}$ sont les opérateurs de courbure de Weyl. Par suite si R est un opérateur de courbure, il admet la décomposition en composantes $O(V)$-irréductibles $R = Z + W + U$. En calculant successivement la trace de R en tant qu'endomorphisme de $\Lambda^2(V)$, puis sa trace de Ricci, on obtient les valeurs suivantes des termes U et Z :

$$U = \frac{u}{2n(n-1)}\, g \bigcirc\!\!\!\!\wedge g\,,$$

$$Z = \frac{1}{n-2}\left(r - \frac{u}{n}\, g\right) \bigcirc\!\!\!\!\wedge g\,,$$

après avoir observé par ailleurs que l'adjoint de l'opérateur $S_o^2\Lambda^2(V)\to S_o^2V$ qui associe à un opérateur de courbure dans $S_o^2\Lambda^2(V)$ sa trace de Ricci est $\frac{1}{n-2}\cdot \circledA g$. Le terme W est ce qui reste de R.

Lorsque $n = 4$ et V orienté, sous l'action du groupe spécial orthogonal $SO(V)$ sur $\mathcal{R}$, $\mathcal{Z}$ reste irréductible tandis que $\mathcal{W}$ se scinde en deux composantes irréductibles $\mathcal{W}^+$ et $\mathcal{W}^-$ correspondant aux valeurs propres $+1$ et -1 de l'opérateur $*$ de Hodge agissant sur les 2-formes par $\langle *\alpha,\beta\rangle = \langle\alpha\wedge\beta, e_1\wedge e_2\wedge e_3\wedge e_4\rangle$, où $(e_i))$ est une base orthonormée orientée de V, α et β deux 2-formes sur V. Alors $\mathcal{W}^\pm = \{W|W\epsilon \quad W\circ * = \pm W\}$. Si V est l'espace tangent en un point à une variété riemannienne orientée (M,g) de dimension 4, on dira que (M,g) est <u>auto-duale</u> (resp. anti-auto-duale) si $\mathcal{W}^- = 0$ (resp. $\mathcal{W}^+ = 0$).

1.10. A partir de maintenant (M,g) désigne une variété close (i.e. compacte, sans bord, connexe), de classe C^∞, munie d'une métrique riemannienne g, fixée. On note v_g l'élément de volume qu'elle définit sur M. La classe conforme de g, i.e. la classe des métriques f^2g, où f est une fonction positive dans $C^\infty(M)$, sera <u>notée</u> γ. Sur chaque espace tangent à (M,g), le champ de tenseurs de courbure R de g et son champ de tenseurs de courbure de Weyl induisent des opérateurs de même nom tandis que le champ de tenseurs de Ricci r donne la trace de Ricci. Nous noterons u la courbure scalaire de g. Les propriétés de décomposition irréductible (1.9) se prolongent au champ de tenseurs de courbure R.

Nous noterons $e : E \to M$ un fibré vectoriel au-dessus de M, E_x sa fibre en x dans M, $C^\infty(E)$ ou $C^\infty_M(E)$ l'espace de ses sections C^∞. Ce fibré vectoriel est supposé associé à un fibré principal P dont le groupe structural est un sous-groupe classique G du groupe linéaire général réel. Soit $\mathfrak{g}$ l'algèbre de Lie de G. On note $\mathfrak{g}_E \to M$ le fibré vectoriel des algèbres de Lie contenu dans Hom(E,E). La fibre $(\mathcal{G}_E)_x$ en x est l'espace des endomorphismes antisymétriques de E_x, qui annulent (en tant que dérivation de ΛE_x) les tenseurs définissant la G-structure. On observera que $C^\infty(\mathcal{G}_E)$ est l'algèbre de jauge de E et que $\mathcal{G}_E$ est isomorphe à $E \times_{Ad} \mathcal{G}$.

Soit ∇ une connexion fixée sur $e : E \to M$. Alors à ∇ est associée une 2-forme de courbure R^∇ à valeurs dans $\mathcal{G}_E$ et donnée sur un couple de vecteurs tangents x,y par

(1.11) $$R^\nabla_{xy} = - [\nabla_x , \nabla_y] + \nabla_{[xy]} \quad .$$

Puisque nous considérons souvent dans la suite des 2-formes à valeurs dans $\mathcal{G}_E$, nous notons $C^\infty(\Lambda^k T^*M, K)$ l'espace des sections C^∞ du fibré vectoriel des k-formes différentielles extérieures sur M à valeurs dans un fibré vectoriel K.

Pour deux 2-formes A et B dans $C^\infty(\Lambda^2 T^*M \quad)$ nous posons

(1.12) $$[A,B](x,y) = \sum_{i=1}^{n} \{[A_{e_i,x} , B_{e_i,y}] - [A_{e_i,y} , B_{e_i,x}]\}$$

où (e_i) est une base orthonormée de TmM, espace tangent en m à M, et x et y deux vecteurs de cet espace. Le [,] à droite se réfère au crochet de Lie dans les fibres de $\mathcal{G}_E$.

2. LAPLACIENS NATURELS ET FORMULES DE WEITZENBÖCK

2.1. Nous commençons cette section par quelques rappels sommaires sur les opérateurs différentiels (o.d.). Pour des détails sur ce sujet on peut voir par exemple [16, chap. IV] et [13].

Un o.d. d'ordre inférieur ou égal à un entier $k \geqslant 0$ sur le fibré vectoriel $e : E \to M$ à valeurs dans le fibré vectoriel $p : F \to M$ est une application linéaire $D : C^\infty(E) \to C^\infty(F)$ telle que pour toute fonction f C^∞ sur M, Df - fD est encore un o.d. mais d'ordre k-1, les o.d. d'ordre zéro étant des homomorphismes de $C^\infty(M)$-modules. En particulier une application linéaire $D : C^\infty(E) \to C^\infty(F)$ telle que pour f dans $C^\infty(M)$ et ω dans $C^\infty(E)$ et $f(x) = 0 = \omega_x$, on ait $D(f\omega)\, x = 0$ est un o.d. d'ordre inférieur ou égal à 1.

2.2. Par exemple la dérivée covariante notée encore ∇ associée à la connexion ∇ sur $e : E \to M$ est un o.d. d'ordre $\leqslant 1$ puisque pour f dans $C^\infty(M)$ et ω dans $C^\infty(E)$

$$\nabla(f\omega) = f\nabla\omega + \omega \otimes df \quad .$$

Il en est de même de la <u>différentielle extérieure</u> d^∇ associée à ∇ et agissant sur $C^\infty(\Lambda T^*M, E)$. Si au lieu de la connexion ∇ elle-même, on considère sur le fibré vectoriel $\Lambda^k TM \otimes E \to M$ la connexion notée encore ∇, produit tensoriel de la connexion de Levi-Civita D sur le fibré tangent et de ∇, l'o.d. d^∇ s'écrit pour ω dans $C^\infty(\Lambda^k T^*M, E)$ et $(x_0, \dots x_k)$ k+1 champs de vecteurs sur M

$$(2.3) \qquad (d^\nabla\omega)(x_0, \dots x_k) = \sum_{i=1}^{k} (-1)^i (\nabla_{x_i}\omega)(x_0, \dots, x_{i-1}, x_{i+1}, \dots, x_k) \; .$$

2.4. Sur $C^\infty(\Lambda^k T^*M, E)$ on définit un produit scalaire global en posant

$$(\omega,\omega') = \int_M \sum_{i_1<\dots<i_k} < \omega(e_{i_1},\dots,e_{i_k}) , \omega'(e_{i_1},\dots,e_{i_k}) > v_g$$

où $<,>$ est un produit scalaire sur les fibres de E et (e_i) une base orthonormée de l'espace tangent en un point de M. L'adjoint (formel) de d^∇ par rapport à $(,)$ est la <u>codifférentielle extérieure</u> δ^∇, donné pour ω dans $C^\infty(\Lambda^{k+1}T^*M, E)$ par

$$(\delta^\nabla\omega)_{x_1,\dots x_k} = - \sum_{i=1}^n (\nabla_{e_i}\omega)(e_i, x_1,\dots, x_k)$$

où $x_1,\dots,x_k$ sont des vecteurs tangents. On vérifie ici encore que δ^∇ est un o.d. d'ordre $\leqslant 1$.

2.5. Soit $D : C^\infty_M(E) \to C^\infty_M(F)$ un o.d. d'ordre $\leqslant k$. Pour tout point x dans M et pour tout vecteur cotangent t dans T^*_xM non nul le <u>symbole</u> de D est l'application linéaire $\sigma_t(D) : E_x \to F_x$ donnée par

$$\sigma_t(D)e = \frac{1}{k!} D(f^k\omega)(x)$$

où f est une fonction C^∞ sur M telle que $f(x) = 0$, $df(x) = t$ et ω une section dans $C^\infty_M(E)$ telle que $\omega(x) = e$. On vérifie facilement que $\sigma_t(\nabla)\cdot = t \otimes \cdot$; $\sigma_t(d^\nabla)\cdot = t \wedge \cdot$.

Si D_1 et D_2 sont deux opérateurs différentiels, $D_1 \circ D_2$ est un o.d. et $\sigma_t(D_1 \circ D_2) = \sigma_t(D_1)\circ \sigma_t(D_2)$. De même si D^* désigne l'adjoint de D pour le produit scalaire global introduit dans 2.4., $\sigma_t(D^*) = (-1)^k (\sigma_t(D))^*$. Ainsi $\sigma_t(\delta^\nabla) = - i_t$, où i_t désigne le produit intérieur par le vecteur t.

Un o.d. D est appelé elliptique si son symbole $\sigma_t(D)$ est un isomorphisme de E_x sur F_x pour tout x dans M et tout t non nul dans T^*_xM.

2.6. Considérons l'opérateur différentiel $D_M : C^\infty_M(E) \to C^\infty_M(F)$ et (N,h) une variété riemannienne. On dit que D_M est un o.d. naturel (ou riemanien) s'il satisfait l'axiome de naturalité suivant:

Pour toute isométrie $\varphi : (M,g) \to (N,h)$ le diagramme

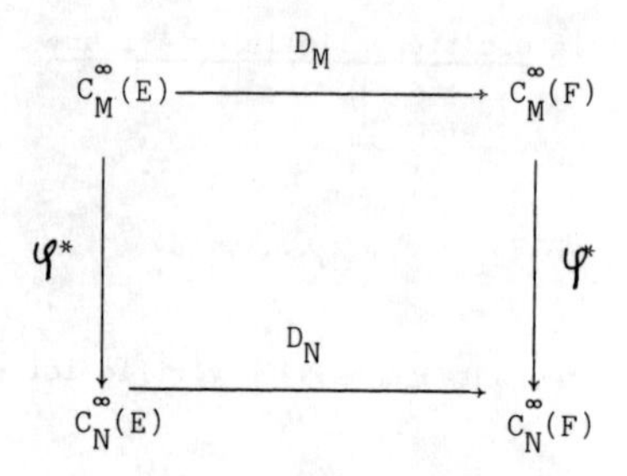

commute. Les opérateurs différentiels mentionnés dans les paragraphes précédents ont cette propriété fonctorielle. Les o.d.n. ont été très étudiés dans les années 70 (voir par exemple [9],[18],[19] pour plus de détails sur le sujet).

Nous nous intéressons ici aux laplaciens naturels, c'est-à-dire aux o.d.n. du second ordre dont le symbole est l'opposé de la métrique riemannienne sur M. Par leur définition même ils sont elliptiques. En outre la différence de deux laplaciens naturels agissant sur les sections C^∞ d'un fibré vectoriel au-dessus de M est nulle ou est un opérateur différentiel d'ordre 0, déterminé par la courbure de la métrique g sur M. L'ensemble des laplaciens naturels est donc un espace affine associé à l'espace vectoriel des endomorphismes de E.

Par exemple si D est la connexion de Levi-Civita sur (M,g), on observe que D^*D est un laplacien naturel. Il est de la famille des laplaciens dits "forts" [3], [4, exposé XVI] qui est un cône de l'espace affine des laplaciens naturels.

2.7. On range sous les termes de <u>formules de Weitzenböck</u> toute expression de comparaison de deux laplaciens forts. Comme nous l'avions noté plus haut la différence entre les deux opérateurs est un terme en courbure de la variété base. C'est donc cette courbure, et aussi l'intérêt recherché qui dicteront le plus souvent les laplaciens forts les mieux adaptés pour la comparaison (pour d'autres commentaires voir [5, §4] et [4, exposé XVI] où l'on trouvera également une série de formules de Weitzenböck écrites de façon intrinsèque sans utiliser des indices).

Pour le théorème d'isolation que nous avons en vue, nous sommes intéressés par la comparaison du laplacien généralisé de Hodge-de Rham $\Delta^\nabla \equiv d^\nabla\delta^\nabla+\delta^\nabla d^\nabla$ et le laplacien fort $\nabla^*\nabla$ (qui est juste l'opposé de la trace du Hessien ∇^2), opérant l'un et l'autre sur les 2-formes sur M à valeurs dans $\mathcal{G}_E$. Pour ω dans $C^\infty(\Lambda^2T^*M, \mathcal{G}_E)$, on a par définition

$$(2.8) \qquad (\nabla^*\nabla\omega) = -\sum_{i=1}^{n} \nabla^2_{e_i,e_i}\omega$$

où (e_i) est une base orthonormée de l'espace tangent en un point à M et $\nabla^2_{x,y} = \nabla_x\nabla_y - \nabla_{D_x y}$.

Alors on a

2.9. <u>Proposition</u>. Pour toute 2-forme ω sur M à valeurs dans

$$(2.10) \qquad \Delta^\nabla\omega = \nabla^*\nabla\omega + \frac{2u}{(n-1)(n-2)}\omega + \frac{n-4}{n-2}\,\omega o(r \circledcirc g) - 2\omega oW - [R^\nabla,\omega].$$

<u>Preuve</u>. Suivant [7], fixons un point m dans M. Soient $(e_i)(i=1,\ldots n)$, x, y une base orthonormée et deux vecteurs de l'espace tangent T_mM. Ils se prolongent en des champs de vecteurs locaux que nous notons encore (e_i)

$(i=1,\ldots n)$, x, y, sur M tels que (e_i) soit un champ de repères orthonormés locaux et $(De_1)(x) = \ldots = (De_n)(x) = 0 = (DX)(m) = (DY)(m)$, où D est la connexion de Levi-Civita sur M. Il suit en particulier que $\nabla_x\nabla_y = \nabla^2_{xy}$. Alors pour ω dans $C^\infty(\Lambda^2T^*M, \mathcal{G}_E)$

$$d^\nabla(\delta^\nabla\omega)(x,y) = \sum_{i=1}^n \{(\nabla^2_{y,e_i}\omega)(e_i,x) - \sum_{i=1}^n (\nabla^2_{x,e_i}\omega)(e_i,y)$$

$$\delta^\nabla(d^\nabla\omega)(x,y) = -\sum_{i=1}^n \nabla_{e_i}(d^\nabla\omega)(e_i,x,y) =$$

$$-\sum_{i=1}^w (\nabla^2_{e_i,e_i}\omega)(x,y) + \sum_{i=1}^n (\nabla^2_{e_i,x}\omega)(e_i,y) - \sum_{i=1}^w (\nabla^2_{e_i,y}\omega)(e_i,x).$$

Il en résulte

$$\Delta^\nabla\omega(x,y) = (\nabla^*\nabla\omega)(x,y) - \sum_{i=1}^n (R^\nabla_{xe_i}\omega)(e_i,y) - \sum_{i=1}^n (R^\nabla_{ye_i}\omega)(e_i,x).$$

Mais R^∇_{xy} est une dérivation tensorielle, i.e.

$$(R^\nabla_{xy}\omega)(u,v) = [R^\nabla_{xy},\omega(u,v)] - \omega(u,R^\nabla_{xy}v) - \omega(R_{xy}u,v),$$

d'où

$$\Delta^\nabla\omega(x,y) = (\nabla^*\nabla\omega)(x,y) + \omega(r(x),y) - \omega(r(y),x) + \sum_{i=1}^n \omega(e_i,R_{ye_i}x)$$

$$-\sum_{i=1}^n \omega(e_i,R_{xe_i}y) + \sum_{i=1}^n \{[R^\nabla_{xe_i},\omega(e_i,y)] - [R^\nabla_{ye_i},\omega(e_i,x)]\}.$$

De la première identité de Bianchi, de (1.12) et de la convention posée en 1.3., il suit que

$$\Delta^\nabla\omega(x,y) = \nabla^*\nabla\omega(x,y) + (\omega\circ(r\wedge g))_{xy} - 2(\omega\circ R)_{xy} - [\overset{\nabla}{R},\omega](x,y),$$

D'où

$$(2.11) \qquad \Delta^{\nabla}\omega = \nabla^{*}\nabla\omega - \omega\circ(2R - r \otimes g) - [R^{\nabla},\omega].$$

En remplaçant dans (2.11) R par ses composantes irréductibles sous l'action du groupe orthogonal, on obtient (2.10) || .

2.12. Remarques

1) On observe sur (2.10) qu'en dimension 4 le terme en courbure donnant $\Delta^{\nabla}\omega - \nabla^{*}\nabla\omega$ ne contient pas la courbure de Ricci. En cette dimension (2.10) se réduit à

$$(2.13) \qquad \Delta^{\nabla}\omega = \nabla^{*}\nabla\omega + \frac{u}{3}\,\omega - 2\omega\circ W - [R^{\nabla},\omega],$$

formule qui sera centrale dans la Section 3.

2) En fait en écrivant que

$$2R - r \otimes g = - \frac{2(n-2)}{n(n-1)}\,\mathrm{Id}_{\Lambda^2 TM} - \frac{n-4}{n-2}\left(r - \frac{u}{n}\,g\right) \otimes g + 2W$$

on remarque que c'est la partie déviante de la courbure de Ricci qui n'intervient pas en dimension 4. Ce fait est général dans les formules de Weitzenböck pour les laplaciens agissant sur les k-formes différentielles extérieures sur les variétés de dimension 2k et peut être exploité pour obtenir des théorèmes d'annulation, c.f. par exemple [6].

3. ISOLATION L^2 DES CHAMPS DE YANG-MILLS

3.1. Nous fixons la dimension de (M,g) égale à 4 et supposons que la courbure scalaire de g est positive. Sur l'espace affine des G-connexions ∇ du fibré vectoriel $e : E \to M$, on définit <u>la fonctionnelle de Yang-Mills</u> par

$$(3.2) \qquad \nabla \mapsto \mathcal{YM}(\nabla) = \frac{1}{2} \int_M |R^\nabla|^2 \, v_g$$

où la norme $|\ |$ se réfère à la métrique g et au produit scalaire Ad-invariant fixé sur $\mathcal{G}$.

Les points critiques de $\mathcal{YM}$ sont appelés les <u>connexions</u> de Yang-Mills et leurs tenseurs de courbures R^∇ les <u>champs de Yang-Mills</u>. En tant que 2-formes de courbure ces tenseurs sont ∇-<u>harmoniques</u> i.e. vérifient $d^\nabla R^\nabla = 0$ et $\delta^\nabla R^\nabla = 0$. En effet la première égalité est la seconde identité de Bianchi $(\nabla_x R^\nabla)_{y\,z} + (\nabla_y R^\nabla)_{zx} + (\nabla_z R^\nabla)_{xy} = 0$ et la deuxième est l'équation d'Euler-Lagrange de (3.2). Pour chaque connexion ∇, le nombre réel $\mathcal{YM}(\nabla)$ ne dépend que de la classe conforme γ de g (et non de la métrique spécifique g dans cette classe); pour d'autres propriétés de la fonctionnelle $\mathcal{YM}$, voir par exemple le livre de M.F. ATIYAH, <u>Geometry of Yang Mills fields</u>, Pisa, Scuola Normale Superiore (Lezioni fermiane) 1979.

3.3. Dans cette situation conforme, nous pouvons relier la fonctionnelle de Yang-Mills à un autre invariant de la classe γ qui est le <u>nombre de</u> Yamabe μ_γ défini pour une variété de dimension n par

$$\mu_\gamma = \inf \{ (\textstyle\int_M u_{\bar g} \, v_{\bar g}) (\int_M v_{\bar g})^{-1+\frac{2}{n}}, \ \bar g \in \gamma \}$$

où $u_{\bar g}$ désigne la courbure scalaire de $\bar g$.

3.4. <u>Lemme</u>. <u>Sur une variété</u> (M,γ) <u>de dimension</u> $n \geqslant 3$,

$$(3.5) \quad \mu_\gamma = \inf \{[\int (A|df|^2 + u_g f^2) v_g] (\int f^N v_g)^{-1+\frac{2}{n}} \quad , f \in C^\infty(M), f > 0 \}$$

où $A = \frac{4(n-1)}{n-2}$ et $N = \frac{2n}{n-2}$.

<u>Preuve</u>: Facile à partir de l'équation donnant $u_{\bar{g}}$ en fonction de u_g [20, (1.8)] ||.

Le nombre de Yamabe est défini pour toute variété (M,g) et réalise la courbure scalaire d'une métrique dans la classe γ : c'est la solution du problème de Yamabe résultant de [1],[12],[17]. Pour d'autres propriétés μ_γ voir [10].

3.6. Utilisant μ_γ on a pu isoler dans la topologie L^2 ([6], th. B) des champs de Yang-Mills auto-duaux et anti-auto-duaux. Cela signifie que il est montré l'existence d'un voisinage L^2 de champs de Yang-Mills minimaux dans lequel il n'y a pas d'autres champs de Yang-Mills sur la variété (M,g) supposée orientée. Notons R_+^∇ et R_-^∇ les deux composantes de la 2-forme de courbure R^∇ sous l'action de l'opérateur $*$ de Hodge. Il est connu que $R_\pm^\nabla$ sont harmoniques si et seulement si R^∇ l'est et qu'ils réalisent des minima absolus de la fonctionnelle de Yang-Mills.

Alors comme généralisation d'un théorème d'isolation L^2 de J. Dodziuk et Min-Oo [8], nous pouvons énoncer

3.7. <u>Théorème</u> [6]

<u>Sur une variété</u> (M,γ) <u>auto-duale avec</u> $\mu_\gamma > 0$, <u>tout champ de Yang-Mills</u> R^∇ <u>sur un</u> SU(2)-<u>fibré est auto-dual</u> (i.e. $R_-^\nabla \equiv 0$) <u>dès que</u>

(3.8) $$48 \int_M |R_-^\nabla|^2 < \mu_\gamma^2 .$$

<u>Si l'inégalité est large, alors soit le champ est auto dual, soit sa partie négative est paralléle</u>.

La preuve résulte de l'inégalité générale suivante, valable pour toute 2-forme ω à valeurs vectoriels et une SU(2)-connexion ∇ sur un fibré au-dessus de (M,γ)

(3.9) $$(\Delta^\nabla \omega,\omega) \geqslant \left[\frac{1}{6}\,\mu_\gamma - \frac{2}{\sqrt{3}}\,\|R_-^\nabla\|_{L^2} - \|W\|_{L^2}\right] \|\omega\|_{L^4}^2$$

On utilise de façon essentielle l'inégalité de Kato (cf. par exemple [2, p. 130]) pour obtenir (3.9). Pour les détails, voir [6].

3.10. <u>Remarque</u>

Le théorème s'applique en particulier à la sphère $(S^4,[can])$, au projectif complexe $\mathbb{C}P^2$ et aux exemples de variétés compactes auto-duales dus à J. POON [16]. Nous avons donc

$$\int_{S^4} |R_-^\nabla|^2 < 8\,\pi^2 \Rightarrow R_-^\nabla = 0 ,$$

puisque $\mu_{[can]} = n(n-1)\,\omega_n^{\frac{2}{n}}$ sur S^n de volume ω_n.

BIBLIOGRAPHIE

[1] AUBIN, T.: Equations différentielles non linéaires et problème de Yamabe concernant la courbure scalaire, J. Math. Pures et Appl. 55, 269-296 (1976).

[2] BERARD, P.H.: Spectral Geometry: Direct and Inverse Problems, Lecture notes in Math. 1207, Springer-Verlag.

[3] BERGER, M. et EBIN, D.: Some decompositions of the space of symmetric tensors on a Riemannian manifold, J. Diff. Geom. 3, 379-392 (1969).

[4] BOURGUIGNON, J.P.: Formules de Weitzenböck en dimension 4. Exposé XVI, in Géométrie riemannienne en dimension 4, Edit. A. Besse, Cédic/Fernand Nathan, Paris.

[5] BOURGUIGNON, J.P.: Les variétés de dimension 4 a signature non nulle dont la courbure est harmonique sont d'Einstein, Inv. Math. 63, 263-286 (1981).

[6] BOURGUIGNON, J.P. et EZIN, J.P.: Nombre de Yamabe et théorèmes d'annulation pour les formes harmoniques de dimension moitié, à paraître.

[7] BOURGUIGNON, J.P. et LAWSON, H.B.: Stability and isolation phenomena for Yang-Mills fields, Com. Math. Phys. 79, 189-230 (1981).

[8] DODZIUK, J. et Min-Oo.: An L_2-isolation theorem for Yang-Mills fields over complete manifolds, Comp. Math. 47, 165-169 (1982).

[9] EPSTEIN, D.B.: Natural tensors on Riemannian Manifolds, J. Diff. Geom. 10, 631-645 (1975).

[10] EZIN, J.P.: Constante de Yamabe et nombres caractéristiques de variétés riemanniennes à courbure scalaire positive, à paraître.

[11] GALLOT, S. et MEYER, D.: Opérateur de courbure et laplacien des formes différentielles d'une variété riemannienne, J. Math. Pures et Appl. 54, 259-284 (1975).

[12] GIL-MEDRANO, O.: On the Yamabe Problem concerning the compact locally conformally flat manifolds, J. Funct. Anal. 66, 42-53 (1986).

[13] GROMOV, M. et LAWSON, H.B.: Positive scalar curvature and the Dirac operator on complete Riemannian manifolds, Publ. Math. I.H.E.S. 58, 295-408 (1983).

[14] KULKARNI, R.S.: On the Bianchi identities, Math. Ann. 199, 175-204 (1972).

[15] PALAIS, R.S.: Seminar on the Atiyah-Singer index theorem, Annals of Math. Studies 57, Princeton Univ. Press (1965).

[16] POON, J.: Compact self-dual manifolds with positive scalar curvature, J. Diff. Geom. 24, 97-132 (1986).

[17] SCHOEN, R.: Conformal deformation of a Riemannian metric to constant curvature, J. Diff. Geom. 20, 479-495 (1984).

[18] STREDDER, P.: Natural differential operators on Riemannian manifolds and representations of the orthogonal and special orthogonal groups, J. Diff. Geom. 10, 647-660 (1975).

[19] TERNG, C.L.: Natural vector bundles and natural differential operators, Am. J. Math. 100, 775-828 (1978).

[20] YAMABE, H.: On a deformation of Riemannian Structures on compact manifolds, Osaka Math. J. 12, 21-37 (1960).

THEORY OF PSEUDO–DIFFERENTIAL OPERATORS OVER C*–ALGEBRAS

Noor Mohammad
Department of Mathematics, Quaid–i–Azam University, Islamabad, Pakistan.

1. INTRODUCTION

Recently, the pseudo–differential operators, which were introduced by Mihlin and Calderon–Zygmund, have been much exploited in the current literature. Several authors (see for instance, [1], [2] and [3]) studied the algebra of these operators for the classical case. The technique of such operators have beautiful geometrical and topological interpretations. Mischenko [5] gave a natural interpretation in terms of C*–algebras of several versions of the theory of pseudo–differential operators on compact smooth manifold, and Mishenko and Famenko [6] derived analogues of the Atiyah–Singer formula for this case. Regarding applications of pseudo–differential operators over C*–algebras, we refer to [4], [5] and [6]. For a general reference about the classical pseudo–differential operators and investigation of their properties, one can see [8].

In the present paper, we study the behaviour of adjoints and composition of pseudo–differential operators in the framework of a C*–algebra. Consequently, we get that the class of pseudo–differential operators of order zero, is a C*–algebra. We also state here L_2–continuity theorem. In this paper, Hilbert C*–modules as defined by Mishenko [5] and Paschke [7] play an important role.

The formulation of such a problem naturally arises to seek an analogue of Atiyah–Bott formula for calculating the index of an elliptic operator on a compact manifold for this case.

In Sec. 2, we give some basic definitions necessary to develop the concept.

In Sec. 3, we define a class of symbols $S^m_{1,0}$, the one introduced by Hormander [1]. Note that, a symbol $a(x,\xi) : A^k \to A^k$ is an A–homomorphism of Hilbert C*–modules. Here A represents an arbitrary C*–algebra and A^k direct sum of k–copies of A. Corresponding to each symbol, we define a pseudo–differential A–operator. Next, we develop the calculus of such operators and observe that each pseudo–differential A–operator admits an adjoint which is again a pseudo–differential A–operator.

In the last Section, we construct pseudo-differential operator on a vector A–bundle and extend routinely the results of the preceding section with certain modifications to this case. Thus we obtain that a class of pseudo–differential operators of order zero acting in the space $L_2(M, E)$ is a C*–algebra, where M is a compact smooth manifold and E is a vector A–bundle.

2. DEFINITION AND PRELIMINARIES

In this section we shall give some basic definitions necessary to develop the theory.

Let X be a bounded open subset of $\mathbf{R}^n$, and let $x = (x_1, \ldots, x_n)$ be the standard coordinates. Let A be an arbitrary C*–algebra with an identity. Denote by $C^\infty(X,A)$ the space of all smooth functons (i.e. of class C^∞) with values in A, by $C_0^\infty(X, A)$ its subspace of functions with compact supports, and by $C_b^\infty(X, A)$ the space of such C^∞–functions whose derivatives are all bounded. Let $\mathcal{S}(\mathbf{R}^n, A)$ denote the space of C^∞–functions $u(x)$ whose derivatives decrease faster than any power of $|x| = \left(\sum_{i=1}^n x_i^2\right)^{1/2}$ as $|x| \to \infty$, i.e. for every multi–indices α, β

$$\sup_{x \in \mathbf{R}^n} |x^\alpha| \, \|\partial_x^\beta u(x)\| < +\infty \tag{2.1}$$

Recall that a multi–index α is an n–tuple $(\alpha_1, \ldots, \alpha_n)$, $x^\alpha = x_1^{\alpha_1} \ldots x_n^{\alpha_n}$ and $\partial_x^\beta = \partial_{x_1}^{\beta_1} \ldots \partial_{x_n}^{\beta_n}$. We shall sometimes write D_x^α for the partial derivative

$$\left(-i\frac{\partial}{\partial x_1}\right)^{\alpha_1} \ldots \left(-i\frac{\partial}{\partial x_n}\right)^{\alpha_n}.$$

For $u \in \mathcal{S}(\mathbf{R}^n, A)$ we define the Fourier transform of u by

$$\hat{u}(\xi) = \int e^{-ix\cdot\xi} u(x)\, dx \tag{2.2}$$

where $\xi \in \mathbf{R}^n$, $x \cdot \xi = x_1\xi_1 + \cdots + x_n\xi_n$ and $dx = dx_1 \ldots dx_n$ is the Lebesque measure on $\mathbf{R}^n$. Here integration is understood over $\mathbf{R}^n$.

The inverse Fourier transform is given by the formula

$$u(x) = \int e^{ix\cdot\xi} \hat{u}(\xi)\, đ\xi, \tag{2.3}$$

where $đ\xi = (2\pi)^{-n} d\xi_1 \ldots d\xi_n$.

We denote by Δ the operator

$$\Delta = -\sum_{i=1}^n \frac{\partial^2}{(\partial x_i)^2}$$

For $u \in C_0^\infty(X, A)$ we put

$$\|u\|_s^2 = \| \int_X ((1+\Delta)^s u^*(x))u(x)dx\|, \tag{2.4}$$

where s is any real number, and denote by $H_0^s(X, A)$ the completion of $C_0^\infty(X, A)$ relative to the Sobolev's norm (2.4).

We denote by $L_2(X, A)$ the space of all measurable functions (i.e. classes) u, for which the integral $\int_X u^*(x)u(x)dx$ converges.

Now we shall define Hilbert C*–module in a similar way as defined by Mishenko [5] and Paschke [7].

Definition 2.1: Let M be a right A–module. By a Hermitian product in the A–module M we mean a sesquilinear functional $<\cdot,\cdot>: M \times M \to A$ satisfying the following conditions:

(a) $\langle m, m\rangle \geq 0 \quad \forall m \in M$;

(b) $\langle m, m\rangle = 0$ only if m = 0;

(c) $\langle m_1, m_2\rangle = \langle m_2, m_1\rangle^* \quad \forall m_1, m_2 \in M$;

(d) $\langle m_1 a, m_2\rangle = a^*\langle m_1, m_2\rangle \quad \forall m_1, m_2 \in M,\ a \in A$,

(e) $\langle m_1, m_2 b\rangle = \langle m_1, m_2\rangle b, \quad \forall\, m_1, m_2 \in M, b \in A$;

An A–module M equipped with a Hermitian product is called a pre–Hilbert C*–module. For a pre–Hilbert C*–module M, define a norm $\|\cdot\|_M$ on M by $\|m\|_M^2 = \|\langle m, m\rangle\|$. The norm in M satisfies the classical inequalities:

(1) $\|m\cdot a\|_M \leq \|m\|_M \cdot \|a\|, \quad \forall\, m \in M, a \in A$;

(2) $\|\langle m_1, m_2\rangle\| \leq \|m_1\|_M \cdot \|m_2\|_M, \quad \forall m_1, m_2 \in M.$

A pre–Hilbert C*–module M which is complete with respect to the norm $\|\cdot\|_M$ is called a Hilbert C*–module.

For a pre–Hilbert C*–module M, we let $M^* = \mathrm{Hom}_A(M, A)$, i.e. the set of bounded A–module maps of M into A. Each $m_1 \in M$ gives rise to a map $\hat{m}_1 \in M^*$ defined by $\hat{m}_1(m_2) = \langle m_1, m_2\rangle$ for $m_2 \in M$. We will call M self–dual if $\widehat{M} = M^*$, i.e. if every map in M^* arises by taking Hermitian product with some fixed $m_1 \in M$. If M is self–dual, M must be complete. But the converse is not true in general; completeness is not enough to ensure self–duality. We shall assume that the homomorphism $\varphi : M \to M^*$ given by the sesquilinear functional is an isomorphism.

One can find numerous examples of such objects in Mishenko [5] and Paschke [7]. For example, the C*–algebra A itself as well as the direct sum A^k of k copies of A, with the Hermitian product given by $\langle x, y\rangle = \sum_{i=1}^{k} x_i y_i^*$, is a Hilbert C*-module. Note that these modules are self-dual.

Denote by $\ell_2(A)$ the space of those sequences $x = (x_1, \ldots, x_n, \ldots), \forall x_n \in A$, for which $\sum_{n=1}^{\infty} x_n^* x_n$ converges in A. We define a Hermitian product in $\ell_2(A)$ by the formula

$$\langle x, y\rangle = \sum_{n=1}^{\infty} x_n^* y_n. \tag{2.5}$$

The convergence of this last series follows from an analogue of Cauchy's inequality for C*-algebras (see [6]):

$$\| \sum_{n=1}^{\infty} x_n^* y_n \|^2 \le \| \sum_{n=1}^{\infty} x_n^* x_n \| \cdot \| \sum_{n=1}^{\infty} y_n^* y_n \|. \tag{2.6}$$

Then $\ell_2(A)$ is a Hilbert C*-module (see [6]). The following result is due to Miscenko and Fomenko [6].

Lemma 2.1: The space $H_0^s(X, A)$ is isomorphic to $\ell_2(A)$ as a Hilbert C*-module.

The Hermitian product in the Hilbert C*-module $H_0^s(X, A)$ is given by

$$\langle u, v\rangle_s = \int_X ((1+\Delta)^s u^*(x)) v(x) dx. \tag{2.7}$$

Note that in $L_2(X, A)$ the Hermitian product is defined as

$$\langle f, g\rangle = \int_X f^*(x) g(x) dx, \tag{2.8}$$

and the norm by

$$\|f\|_{L_2}^2 = \| \int f^*(x) f(x) dx \|. \tag{2.9}$$

For s = 0, $H_0^s(X, A)$ equals $L_2(X, A)$.

Now consider the space $L_2(X, A^k)$. Then as a consequence of Lemma (2.1) and together with Lemma (1.2) in [6] (see also p. 107 therein), we get the following:

Lemma 2.2 $L_2(X, A^k)$ is a Hilbert C*-module and is isomorphic to $\ell_2(A^k)$.

One pleasant property of self-dual Hilbert C*-module is that every bounded module map between two such has an adjoint. The following result is proved in much the same way as the corresponding fact about Hilbert spaces.

Proposition 2.3 Let M be a self-dual Hilbert C*-module, and $T : M \to M$ a bounded module map. Then there is a bounded module map $T^* : M \to M$ such that

$$\langle Tm_1, m_2\rangle = \langle m_1, T^* m_2\rangle, \quad \forall m_1, m_2 \in M.$$

Consider again the self-dual Hilbert C*-module A^k. Let End(k,A) denote the algebra of matrices of order k, with entries in A. Then End(k,A) becomes a C*-algebra relative to the norm:

$$\|T\| = \sup_{\|a\| \le 1} \|Ta\|,$$

where $a = (a_1, \ldots a_k) \in A^k$ and $\|a\| = \| \sum_{i=1}^{k} a_i^* a_i \|^{1/2}$.

3. ALGEBRA OF PSEUDO-DIFFERENTIAL OPERATORS OVER C*-ALGEBRAS

In this section we shall define the classes of symbols $S^m_{\rho,\delta}$ (for the special case when $\rho = 1$, $\delta = 0$) in a way analogous to the one introduced in Hormander [1]. Also such a subclass $S^m_{1,0}$ had been defined by Kohn and Nihenberg in [2] for the classical case. Pseudo-differential operators over C*-algebras are defined through them, in almost the same way as by Miscenko and Fomenko in [6]. We investigate the behaviour of adjoints and products of such operators and their continuity on $L_2(\mathbf{R}^n, A^k)$ spaces

Definition 3.1: Let X be a bounded open subset of $\mathbf{R}^n$ and let m be any real number. Denote by $S^m_{1,0}(X \times \mathbf{R}^n, \mathrm{End}(k, A))$ the space of such functions $a(x, \xi)$ with the following properties:

(a) $a(x, \xi)$ have compact supports in the x variable;

(b) for all multi indices α, β, there exists a constant $C_{\alpha,\beta}$ such that

$$\|\partial_\xi^\alpha \partial_x^\beta a(x, \xi)\| \le C_{\alpha\beta} \langle \xi \rangle^{m-|\alpha|}, \tag{3.1}$$

where $x \in X$, $\xi \in \mathbf{R}^n$ and $\langle \xi \rangle = \left(1 + \sum_{i+1}^{n} \xi_i^2\right)^{1/2}$. We recall that the support of $a(x, \xi)$ in the variable x is the closure of the set of points $x \in X$ for which there exists some $\xi \in \mathbf{R}^n$ such that $a(x, \xi) \neq 0$, and it will be denoted by supp $a(x, \xi)$.

We shall use $S^m_{1,0}$ to denote the symbol class when the context is clear. Note that the symbol a is an operator-valued function $a = a(x, \xi) : A^k \to A^k$. Such symbols lead to operators P defined by

$$Pu(x) = \int\int e^{i(x-y)\cdot\xi} a(x, \xi) u(y) dy \, đ\xi, \tag{3.2}$$

where $u \in C_0^\infty(x, A^k)$.

Recall that in (3.2) we integrate partially with respect to y first and then with respect to ξ. We can rewrite the integral (3.2) as

$$Pu(x) = \int e^{ix\cdot\xi} a(x, \xi) \hat{u}(\xi) \, đ\xi, \tag{3.3}$$

which is absolutely convergent.

Definition 3.2: The operator P defined by (3.2) will be called a pseudo–differential A–operator of order m, if $a(x,\xi) \in S^m_{1,0}$.

We denote by L^m the set of pseudo–differential A–operators of order m and set $L^{-\infty} = \cap_m L^m$.

Theorem 3.3: Every pseudo–differential A–operator of order m defines a continuous linear mapping of $C_0^\infty(X, A^k)$ into itself which extends as a continuous linear mapping of $C_b^\infty(X, A^k)$ into itself.

Proof: If $a(x,\xi) \in S^m_{1,0}$, $u \in C_0^\infty(X, A^k)$, then the integral $Pu(x) = \int e^{ix\cdot\xi} a(x,\xi)\hat{u}(\xi)\, đ\xi$ is absolutely convergent, and one can differentiate under the integral sign, obtaining always absolutely convergent integrals. Since $a(x,\xi)$ have compact supports in the x–variable, therefore $(Pu)(x) = 0$, whenever $x \notin$ supp $a(x,\xi)$. Thus supp $(Pu) \subset$ supp $a(x,\xi)$ and the desired result follows immediately.

Now let $u \in C_b^\infty(X, A^k)$, then after suitable substitution the integral $Pu(x) = \int\int e^{i(x-y)\cdot\xi} a(x,\xi) u(y) dy\, đ\xi$ can be expressed as

$$Pu(x) = \int\int \left\{ (1+\Delta_y)^{N_1} \frac{e^{i(x-y)\cdot\xi}}{(1+|x-y|^2)^{N_2}} \right\} \times \left\{ (1+\Delta_\xi)^{N_2} \frac{a(x,\xi)}{(1+|\xi|^2)^{N_1}} \right\} u(y) dy\, đ\xi$$

$$= \int\int \frac{e^{i(x-y)\cdot\xi}}{(1+|x-y|^2)^{N_2}} \left\{ (1+\Delta_\xi)^{N_2} \frac{a(x,\xi)}{(1+|\xi|^2)^{N_1}} \right\} \{ (1+\Delta_y)^{N_1} u(y) \} dy\, đ\xi$$

where N_1 and N_2 are natural numbers such that $N_1 > m + \frac{n}{2}$ and $N_2 > \frac{n}{2}$.

The last integral is absolutely convergent, and thus differentiating under the integral sign, the continuity of P follows, which completes the proof.

Remark: By means of Theorem (3.3), we can easily extend the operator $P \in L^m$ as a continuous map of $\mathcal{S}(\mathbf{R}^n, A^k)$ into $\mathcal{S}(\mathbf{R}^n, A^k)$.

Note that $a(x,\xi)$ vanishes for every $\xi \in \mathbf{R}^n$, whenever x is outside supp $a(x,\xi)$. This justifies the consideration of the operator P on $\mathcal{S}(\mathbf{R}^n, A^k)$.

Viewing $L_2(\mathbf{R}^n, A^k)$ as a Hilbert C*–module, we now want to prove the L_2–continuity theorem. Consider the pseudo–differential A–operator P which maps $C_0^\infty(\mathbf{R}^n, A^k)$ into $C_0^\infty(\mathbf{R}^n, A^k)$. Then we have

Theorem 3.4: Every pseudo–differential A–operator of order zero can be extended to a bounded map of $L_2(\mathbf{R}^n, A^k)$ into $L_2(\mathbf{R}^n, A^k)$, i.e. there exists a constant $C > 0$ such that

$$\|Pu\|_{L_2} \le C\|u\|_{L_2}, \quad u \in C_0^\infty(\mathbf{R}^n, A^k). \tag{3.4}$$

Thus pseudo–differential A–operator is an A–homomorphism of Hilbert C*–modules.

We recall that L_2-norm is defined as

$$\|u\|_{L_2}^2 = \| \int (u(x), u(x)) dx \|, \quad u = (u_1, \ldots, u_k) \in L_2(\mathbf{R}^n, A^k)$$

where $(u(x), u(x))$ denotes Hermitian product in A^k.

The proof of this theorem mainly follows arguments similar to those in [6]. Hence, we omit the proof.

Next we consider sums of operators of decreasing orders (analogous to terms of different orders for pseudo-differential A-operators).

Theorem 3.5: Suppose $a_j \in S_{1,0}^{m_j}$, $m_j \downarrow -\infty$. Then there exists $a \in S_{1,0}^{m_0}$ such that for all $N > 0$

$$a - \sum_{j=0}^{N-1} a_j \in S_{1,0}^{m_N} \tag{3.5}$$

We shall write $a \sim \sum a_j$ to signify that (3.5) holds for each N.

This theorem is proved on the same lines as in the case of classical pseudo-differential operators.

We now turn to more interesting properties of the operators, that they are closed under composition and involution (i.e. adjoint), and may be invariantly defined on a manifold. First we consider the adjoint of a pseudo-differential A-operator.

Theorem 3.6: If P is a pseudo-differential A-operator of order zero on $L_2(\mathbf{R}^n, A^k)$, then there exists a pseudo-differential A-operator P^* such that

$$\langle Pu, v\rangle = \langle u, P^* v\rangle$$

$$u, v \in C_0^\infty(\mathbf{R}^n, A^k). \tag{3.6}$$

Proof: Let P be a pseudo-differential A-operator of order zero given by

$$Pu(x) = \int\int e^{i(x-y)\cdot\xi} a(x,\xi) u(y) dy\, đ\xi.$$

where $a(x,\xi) \in S_{1,0}^0$. Then by the relation (3.6) we have

$$\langle Pu, v\rangle = \int\int\int e^{-i(x-y)\cdot\xi} u^*(y) a^\#(x,\xi) v(x) dy\, đ\xi dx.$$

$$= \int u^*(y) \left\{ \int\int e^{i(y-x)} a^\#(x,\xi) v(x) dx\, đ\xi \right\} dy,$$

where $a^{\#}(x,\xi) = (a^*_{ji}(x,\xi))$, the matrix of symbol $a(x,\xi)$ being $a_{ij}(x,\xi)$. This gives

$$P^*v(x) = \int\int e^{i(x-y)} a^{\#}(y,\xi) v(y) dy\, d\xi, \tag{3.7}$$

which is obviously a pseudo–differential A–operator. We also remark that (3.6) determines P^* uniquely. This proves the theorem.

•

Note that $a^{\#}(y,\xi)$ is not a symbol of the operator P^*. As a matter of fact, the symbol of the operator P^* has the following asymptotic expansion.

Theorem 3.7: If P is a pseudo–differential A–operator with symbol $a(x,\xi)$ and P^* its adjoint operator as given above, then the symbol of P^* is defined by

$$a^*(x,\xi) \sim \sum_\alpha \frac{1}{\alpha!} \partial^\alpha_\xi D^\alpha_x a^{\#}(x,\xi) \tag{3.8}$$

Next we shall study the composition of pseudo–differential A–operators P and Q.

Theorem 3.8: Let P and Q be pseudo–differential A–operators with symbols $a(x,\xi) \in S^{m_1}_{1,0}$ and $b(x,\xi) \in S^{m_2}_{1,0}$, respectively. Then $R = QP$ is a pseudo–differential A–operator with symbol $C(x,\xi) \in S^{m_1+m_2}_{1,0}$ and one has the analogue of Leibniz's formula:

$$C(x,\xi) \sim \sum_\alpha \frac{1}{\alpha!} \partial^\alpha_\xi b(x,\xi) \cdot D^\alpha_x\, a(x,\xi). \tag{3.9}$$

Theorems (3.7) and (3.8) are proved in a similar way as the corresponding assertions in case of classical pseudo–differential operators, provided we take into consideration the standard estimates while applying to a C*–algebra.

From Theorems (3.7) and (3.8) we get

Corollary 3.9: The class $\mathcal{L}$ of pseudo–differential A–operators of order zero acting on the space $L_2(\mathbf{R}^n, A^k)$ is an algebra with involution (as defined in Theorem 3.6).

As we have remarked above these pseudo–differential A–operators are A–homomorphisms of Hilbert C*–modules which admit bounded adjoint A–homomorphisms by virtue of Theorems (3.6) and (3.4): We also observe that with respect to the operator norm

$$\|P\| = \sup_{\|u\|\le 1} \|Pu\|, \quad u \in L_2(\mathbf{R}^n, A^k), \tag{3.10}$$

where $\|\cdot\|$ denotes L_2–norm, the space $\mathcal{L}$ is complete.

We can consider the operator $P \in End^*_A(\ell_2(A^k))$, where $End^*_A(\ell_2(A^k))$ denotes the space of A–homomorphisms which admit adjoint. The fact that $\mathcal{L}$ is a C*–algebra can be shown as follows. To this end, it remains to show that $\|P^*P\| = \|P\|^2$. We can write

$$\|P\| = \sup_{\|u\|\leq 1,\ \|v\|\leq 1} \|\langle Pu, v\rangle\|, u, v \in L_2(\mathbf{R}^n, A^k).$$

In fact,

$$\|P\|^2 = \sup_{\|u\|\leq 1} \|\langle Pu, Pu\rangle\| \leq \sup_{\|u\|\leq 1} \left(\sup_{\|v\|=\|Pu\|} \|\langle Pu, v\rangle\| \right)$$

$$= \sup_{\|u\|\leq 1} \left(\|Pu\| \cdot \sup_{\|v\|\leq 1} \|\langle Pu, v\rangle\| \right)$$

$$\leq \|P\| \cdot \sup_{\|u\|\leq 1,\ \|v\|\leq 1} \|\langle Pu, v\rangle\|.$$

Thus $\|P^*\| = \|P\|$.

Also we have

$$\|P^*P\| = \sup_{\|u\|\leq 1,\ \|v\|\leq 1} \|\langle P^*Pu, v\rangle\| = \sup_{\|u\|\leq 1, \|v\|\leq 1} \|\langle Pu, Pv\rangle\|$$

$$\geq \sup_{\|u\|\leq 1} \|\langle Pu, Pu\rangle\| = \|P\|^2.$$

But on the other hand,

$$\|P^*P\| = \sup_{\|u\|\leq 1} \|P^*Pu\| \leq \|P^*\| \sup_{\|u\|\leq 1} \|Pu\|$$

$$= \|P^*\| \cdot \|P\| = \|P\|^2.$$

Hence $\|P^*P\| = \|P\|^2$.

Thus we have:

Theorem 3.10: The class $\mathcal{L}$ of pseudo–differential A–operators of order zero acting on the space $L_2(\mathbf{R}^n, A^k)$ is a C*–algebra.

4. PSEUDO–DIFFERENTIAL OPERATORS ON VECTOR A–BUNDLES

This section is devoted to the trivial transference of all the concepts in the theory of pseudo–differential operators on compact manifolds to the case of vector bundles over C*–algebras. We have already treated the local situation of trivial A–bundles in the previous section. First we define pseudo–differential A–operator on compact manifold as defined in [6], and then extend the results described in the preceding section to this case.

Let A be a C*–algebra.

Definition 4.1: By a vector A–bundle we shall mean a locally trivial fiber bundle $\xi = (E, p, M, F, G)$, where E is a total space, M is the base of the fiber bundle, $p : E \to M$ is a projection, F is a fiber of the fiber bundle which is finitely generated projective A-module, and G is the structural group equal to $Aut_A(F)$, the group of A–automorphisms of Hilbert C*–module F. If F is a free k–dimensional Hilbert C*–module, then $Aut_A(F)$ is the same group of invertible k–th order matrices with coefficients in A.

Let the base of the fiber bundle be a smooth compact manifold M, then without loss of generality we can suppose that the gluing functions of the bundle are smooth functions (of class C^∞). We denote by $\Gamma(\xi)$ the space of continuous sections of the bundle ξ. Each A–bundle ξ admits a fiber Hermitian product with values in A. If $\Gamma^\infty(\xi)$ denotes the space of smooth sections (of class C^∞), then one can choose the fiber Hermitian product to be smooth, i.e. for any sections $u_1, u_2 \in \Gamma^\infty(\xi)$ the function $\langle u_1, u_2 \rangle \in C^\infty(M, A)$, where $C^\infty(M, A)$ is the space of smooth functions (of class C^∞) on M. For further details of vector A–bundles we refers to [6] and the references therein.

Let $\{X_\alpha\}$ be an atlas of charts of M, $\{\varphi_\alpha\}$ a smooth partition of unity subordinates to $\{X_\alpha\}$, and ψ_α functions such that supp $\psi_\alpha \subset X_\alpha$, $\psi_\alpha|\text{supp}\,\varphi_\alpha \equiv 1$. If $u \in \Gamma^\infty(\xi)$, then in each chart the cross–section $\varphi_\alpha u$ is a smooth function of the coordinates with values in F, where F is a finitely generated projective Hilbert C*–module.

Let E be a trivial A–bundle on the domain X_α with fiber F. We define Sobolev norms in the space $\Gamma_0^\infty(X_\alpha, E)$ of section with compact support by

$$\|u\|_s^2 = \| \int_{X_\alpha} \langle (1+\Delta)^s u(x), u(x) \rangle dx \|. \tag{4.1}$$

The completion of the space relative to the Sobolev norm is denoted by $H_0^s(X_\alpha, E)$. Then it follows trivially from Lemma (3.1) in [6] that Hilbert C*–module $H_0^s(X_\alpha, E)$ is isomorphic to $\ell_2(F)$.

We defined pseudo–differential A–operator for the local situation for trivial A–bundle in the preceding section. Let $\pi : T^*X \to X$, $X \in \{X_\alpha\}$, be the natural projection of a cotangent bundle. We consider the A–homomorphism of the bundles $a : \pi^*(E) \to \pi^*(E)$ as a family of A–homomorphisms $a(x, \xi) : F \to F$ parametrized by points of the cotangent bundle $(x, \xi) \in T^*X$, and satisfying the inequality of the kind (3.1). We call the A–homomorphism $a(x, \xi)$ the symbol of that operator P which is defined as (3.2). We suppose here that $a(x, \xi)$ is compact in the variable x, i.e. $\pi(\text{supp} a) \subset X$.

Now we put

$$[Pu](x) = \sum_\alpha [P_\alpha(\varphi_\alpha u)](x). \tag{4.2}$$

where $u \in \Gamma^\infty(M, E)$ and P_α is the pseudo–differential A–operator defined by (3.2) in the chart X_α by means of the symbol $a_\alpha(x, \xi) = a(x, \xi)\psi_\alpha(x)$. The operator P defined by

(4.2) is not unique of course, but depends on the choice of the functions φ_α and ψ_α and the local coordinates in the charts X_α.

Using the partition of unity φ_α, we define Sobolev norms in the spaces of sections $\Gamma^\infty(M,E)$. Let $u_1, u_2 \in \Gamma^\infty(M,E)$ be arbitrary sections. We put

$$\langle u_1, u_2\rangle_s = \sum_\alpha \int_{X_\alpha} ((1+\Delta_\alpha)^s \varphi_\alpha(x) u_1(x), \quad \varphi_\alpha(x) u_2(x)) dx,$$

$$\|u\|_s^2 = \|\langle u, u\rangle_s\|. \tag{4.3}$$

The completion of the space $\Gamma^\infty(M,E)$ relative to the Sobolev norms (4.3) will be denoted by $H^s(M,E)$.

Obviously $H^0(M,E)$ is isomorphic to $\ell_2(F)$, the direct sum in the module $\ell_2(A) \times \ldots \times \ell_2(A)$. We note that $H^0(M,E)$ equals $L_2(M,E)$.

In general, if an operator $P : H^s(M,E) \to H^{s-m}(M,E)$ is bounded for all s, then we shall say that the operator P is of order m.

We now formulate some useful properties of pseudo–differential A–operators. Proofs and also further details of these operators can be found in [5] and [6].

Theorem 4.2: Any two Sobolev norms of order s in $\Gamma^\infty(M,E)$ are equivalent.

Theorem 4.3: The pseudo–differential A–operator

$$P : H^s(M,E) \to H^{s-m}(M,E)$$

of order m defined by (4.2) is bounded in the Sobolev norms.

Theorem 4.4: When the functions $\varphi_\alpha, \psi_\alpha$ and the local coordinates are changed in the definition (4.2), the operator P is changed by an addend of lower order.

Theorem 4.5: Let $a_1 : \pi^*(E) \to \pi^*(E)$ and $a_2 : \pi^*(E) \to \pi^*(E)$ be symbols of the pseudo–differential A–operators P_1 and P_2 respectively. Then the operators P_3 and P_2P_1, where $a_3 = a_2 a_1$ (i.e. composition of symbols), differ by an operator of lower order.

The results of the preceding section can also be extended routinely with the obvious modifications to this case. Namely we have

Theorem 4.6: Let P be a pseudo–differential A–operator of order zero, then it can be extended to a bounded operator: $L_2(M,E) \to L_2(M,E)$.

Theorem 4.7: If P is a pseudo–differential A–operator of order zero, then it admits an adjoint operator P^* which is also a pseudo–differential A–operator.

As a consequence of theorems (4.6) and (4.7) together with theorem (3.10), we immediately have

Theorem 4.8: The class of pseudo–differential A–operators of order zero acting in the space $L_2(M, E)$ is a C*–algebra.

REFERENCES

[1] L. Hörmander, *Pseudo-differential operators and hypoelleptic equations*, in Proc.Sym. Pure Math. **10**, 138–183 (1968).

[2] J.J. Kohn and L. Nirenberg, *An algebra of pseudo-differential operators*, Comm.Pure Appl.Math. **18**, 269–305 (1965).

[3] H. Kumano–go, *Algebra of pseudo-differential operators*, J.Fac.Sci.University of Tokyo **17**, 31–50 (1970).

[4] A.S. Mishenko, *The theory of elliptic operators over C*-algebras*, Dokl.Akad.Nauk SSSR **239**, 1289–1291 (1978);
English translation in Soviet Math.Dokl. **19**, 512–515 (1978).

[5] A.S. Mishenko, *Banach algebras, pseudo-differential operators and their applications to K-theory*, Uspekhi Mat. Nauk **34:6**, 67–79 (1979);
English translation in Russian Math. Surveys **34:6**, 77–91 (1979).

[6] A.S. Mishenko and A.T. Fomenko, Math. USSR Izvestija, Vol. 15, No. 1 (1980).

[7] W.L. Paschke, *Innerproduct modules over B*-algebra*, Trans.Amer.Math.Soc. **182**, 443–468 (1973).

[8] M.E. Taylor, *Pseudo-differential operators*, (Princeton University Press, New Jersey, 1981).

INTRODUCTION AUX REPRESENTATIONS DES GROUPES COMPACTS

Saliou TOURE

Institut de la Recherche Mathematique,

Université Nationale de la Côte d'Ivoire, Abidjan 08, Côte d'Ivoire.

Dans l'exposé qui suit, nous allons donner une brève introduction à la théorie des représentations des groupes compacts baseé sur [2]. Mais avant d'y arriver, nous donnerons quelques définitions générales.

Dans tout le paragraphe 1, G désignera un groupe topologique localement compact. Il existe alors sur G une mesure de Haar à gauche μ qui sera souvent notée dx.

§1.- DEFINITION D'UNE REPRESENTATION

Définition 1.1.- Soient G un groupe topologique et E un espace de Banach. On appelle représentation continue de G dans E, un homomorphisme $\pi : g \longmapsto \pi(g)$ de G dans le groupe Aut(E) des opérateurs continus inversibles de E tel que la fonction $g \longmapsto \pi(g)v$ soit continue pour tout $v \in E$.

L'espace E s'appelle l'espace de la représentation π. Nous le noterons souvent E_π. La dimension de E_π s'appelle la dimension de π et se note $\dim(\pi)$.

On dit qu'un sous-espace vectoriel X de E est invariant ou est stable par une représentation π de G dans E, si $\pi(g)v \in X$ pour tout $g \in G$ et pour tout $v \in X$.

Une représentation π d'un groupe topologique G dans un espace de Banach E est dite irréductible si $\{o\}$ et E sont les seuls sous-espaces vectoriels fermés de E qui scient invariants par les opérateurs $\pi(g)$, $g \in G$.

Définition 1.2.- Deux représentations π et π' du groupe topologique G dans les espaces de Banach E et F sont dites équivalentes, et on écrit $\pi \simeq \pi'$, <u>s'il existe un isomorphisme</u> T <u>de</u> E <u>sur</u> F <u>tel que</u> $T \circ \pi(g) = \pi'(g) \circ T$ <u>pour tout</u> $g \in G$.

<u>Définition</u> 1.3.- <u>Soient</u> G <u>un groupe topologique et</u> H <u>un espace hilbertien</u>. <u>On appelle représentation unitaire continue de</u> G <u>dans</u> H, <u>un homomorphisme</u> $\pi : g \longmapsto \pi(g)$ <u>de</u> G <u>dans le groupe des opérateurs unitaires de</u> H, <u>tel que la fonction</u> $g \longmapsto \pi(g)u$ <u>soit continue pour</u> tout $u \in H$.

Dans ce cas, on a

$$\pi(g)^* = \pi(g)^{-1} = \pi(g^{-1})$$

où $\pi(g)^*$ désigne l'adjoint de l'opérateur $\pi(g)$ et $\pi(g)^{-1}$ son inverse.

L'ensemble $\hat{G}$ des classes d'équivalence de représentations unitaires continues irréductibles de G s'appelle le <u>dual</u> de G.

On a souvent besoin de la notion de <u>somme directe hilbertienne</u> d'une famille de représentations unitaires.

Soit $(\pi_i)_{i\in I}$ une famille de représentations unitaires d'un groupe G dans des espaces hilbertiens $(H_i)_{i\in I}$. On pose

$$H = \bigoplus_{i\in I} H_i = \{u = (u_i)_{i\in I} : \sum_{i\in I} \|u_i\|^2 = \|u\|^2 < \infty\}$$

et pour $g \in G$

$$\pi(g)\left(\sum_{i\in I} u_i\right) = \sum_{i\in I} \pi_i(g)u_i.$$

On vérifie facilement que la représentation π est bien définie et qu'elle est unitaire. On la note $\pi = \hat{\bigoplus}_{i\in I}\pi_i$ et on l'appelle la <u>somme directe hilbertienne</u> des représentations π_i.

Si π est une représentation unitaire du groupe G dans l'espace hilbertien H, on a une caractérisation simple des sous-espaces invariants :

<u>Théorème</u> 1.4.- <u>Soit</u> π <u>une représentation unitaire continue d'un groupe</u> G <u>dans un espace hilbertien</u> H_π. <u>Alors pour qu'un sous-espace</u>

vectoriel fermé H_1 de H_π soit invariant par π, il faut et il suffit que le projecteur P sur H_1 commute avec $\pi(g)$ pour tout $g \in G$.

Démonstration.- Si H_1 est invariant par π, le sous-espace orthogonal $H_1^{\perp}$ est aussi invariant par π car si $v \in H_1^{\perp}$, on a pour tout $g \in G$ et pour tout $u \in H_1$,

$$(\pi(g)v|u) = (v|\pi(g)^{*}u) = (v|\pi(g^{-1})u) = 0.$$

Comme $H = H_1 \oplus H_1^{\perp}$, tout $u \in H$ s'écrit de facon unique $u = u_1 + u_2$ avec $u_1 \in H_1$ et $u_2 \in H_1^{\perp}$. D'où

$$\pi(g)u = \pi(g)u_1 + \pi(g)u_2$$

et par suite

$$(P \circ \pi(g))(u) = \pi(g)u_1 = (\pi(g) \circ P)(u)$$

puisque $\pi(g)u_1 \in H_1$ et $\pi(g)u_2 \in H_1^{\perp}$.

Donc

$$P \circ \pi(g) = \pi(g) \circ P$$

pour tout $g \in G$.

Réciproquement, si $P \circ \pi(g) = \pi(g) \circ P$ pour tout $g \in G$ et si $u \in H_1$, alors $P(\pi(g)u) = \pi(g)(P(u) = \pi(g)u$, ce qui montre que $\pi(g)u$ est dans H_1, i.e. H_1 est invariant par π.

Théorème 1.5.- Soient π et π' deux représentations unitaires irréductibles de dimension finie d'un groupe G dans des espaces de Hilbert H_π et $H_{\pi'}$. Si $T : H_\pi \longrightarrow H_{\pi'}$, est un opérateur borné tel que $\pi(g) \circ T = T \circ \pi'(g)$ pour tout $g \in G$, alors ou bien $T \equiv 0$, ou bien T est un isomorphisme de H_π sur $H_{\pi'}$.

Démonstration.- Comme $T \circ \pi(g) = \pi'(g) \circ T$ pour tout $g \in G$, $Im(T)$ et $Ker(T)$ sont invariants par π' et π respectivement. En vertu de l'irréductibilité de π et π' on a

$$Im(T) = \{o\} \text{ ou } Im(T) = H_{\pi'},$$

et

$$\mathrm{Ker}(T) = \{o\} \text{ ou } \mathrm{Ker}(T) = H_\pi.$$

Si $\mathrm{Im}(T) = \{o\}$ ou si $\mathrm{Ker}(T) = H_{\pi'}$, alors $T \equiv 0$. Dans le cas contraire, $\mathrm{Im}(T) = H_{\pi'}$, et $\mathrm{Ker}(T) = \{o\}$, c'est-à-dire T est un isomorphisme de H_π sur $H_{\pi'}$.

Définition 1.6.- *Soit* $\mathcal{L}(H)$ *l'algèbre des opérateurs continus dans un espace hilbertien* H *et soit* $M \subset \mathcal{L}(H)$. *Le commutant* M' *de* M *est* l'ensemble

$$M' = \{T \in \mathcal{L}(H) \ : \ S \circ T = T \circ S \quad \forall S \in M\}.$$

Théorème 1.7 (Lemme de Schur).- *Soit* π *une représentation unitaire continue de dimension finie d'un groupe* G *et soit* $M = \{\pi(g) : g \in G\}$. *Alors* π *est irréductible si et seulement si le commutant* M' *de* M *est l'ensemble des opérateurs scalaires.*

Démonstration.- Soit E un sous-espace vectoriel invariant de l'espace H_π de la représentation π et soit P le projecteur sur E. Alors d'après le Théorème 1.4, on a $P \in M'$. Donc si M' est formé d'opérateurs scalaires, on a $P = \lambda I$, où $\lambda \in \mathbb{C}$ et où I désigne l'application identique de H_π.

Comme $P^2 = P$, on a $\lambda = 0$ ou $\lambda = 1$, i.e. $P = 0$ ou $P = I$. On en déduit $E = \{o\}$ ou $E = H_\pi$ et π est irréductible.

Réciproquement, supposons que π soit irréductible. Alors d'après le Théorème 1.5, tout élément de M' est soit nul soit inversible. Soit $T \in M'$ et soit λ une valeur propre de T. L'opérateur $T - \lambda I$ appartient à M' et n'est pas inversible, donc $T - \lambda I = 0$. Comme T est un élément arbitraire de M', on a $M' = \mathbb{C}.I$.

Remarque .- Le Théorème 1.7 reste vrai même si la dimension de π est infinie mais la démonstration utilise alors l'analyse spectrale d'un opérateur hermitien.

Corollaire 1.8.- *Toute représentation unitaire irréductible d'un groupe abélien* G *est de dimension un.*

Démonstration.- Soit π une représentation unitaire irréductible du groupe abélien G dans l'espace hilbertien H_π et soit $M = \{\pi(g) : g \in G\}$.

Comme G est abélien, on a $M \subset M'$. π étant irréductible, M' est formé d'opérateurs scalaires, il en est donc de même de M et tout sous-espace vectoriel de H_π est invariant par π. Si on avait $\dim(H_\pi) > 1$, il existerait des sous-espaces vectoriels fermés différents de $\{o\}$ et H_π qui soient invariants par π (par exemple les sous-espaces vectoriels de dimension un). Donc $\dim(H_\pi) = 1$.

Définition 1.9.- Soit π une représentation unitaire continue d'un groupe topologique G dans un espace hilbertien H_π. On dit qu'un vecteur $v \in H_\pi$ est cyclique pour la représentation π si le plus petit sous-espace invariant contenant v est H_π tout entier.

Une représentation est dite cyclique si elle possède au moins un vecteur cyclique.

On démontre que toute représentation unitaire d'un groupe topologique G est somme directe de représentations cycliques.

§2.- REPRESENTATIONS UNITAIRES DES GROUPES COMPACTS

Dans ce paragraphe, G désignera un groupe topologique compact. Un tel groupe est unimodulaire et il existe sur G une mesure de Haar μ, souvent notée dg, normalisée par la condition

$$\int_G d\mu(g) = 1.$$

Théorème 2.1.- Soit π une représentation continue du groupe compact G dans un espace hilbertien H_π. Alors il existe sur H_π un produit scalaire définissant une norme équivalente à la norme donnée et tel que π soit unitaire pour ce nouveau produit scalaire.

Démonstration.- Soit $(.|.)$ le produit scalaire initial et posons

$$\langle \xi, \eta \rangle = \int_G (\pi(g)\xi \,|\, \pi(g)\eta)\, dg.$$

On vérifie facilement que $(\xi, \eta) \longmapsto \langle \xi, \eta \rangle$ est une forme hermitienne positive sur H_π. Cette forme est non dégénérée car si $\langle \xi, \xi \rangle = o$, alors $(\pi(g)\xi \,|\, \pi(g)\xi) = o$ pour presque tout $g \in G$. En particulier pour $g = e$ (élément neutre de G), on a $(\xi|\xi) = o$ soit $\xi = o$.

Montrons que la représentation π est unitaire par rapport à ce nouveau produit scalaire. On a

$$<\pi(s)\xi,\pi(s)\eta> = \int_G (\pi(g)\pi(s)\xi|\pi(g)\pi(s)\eta)dg$$

$$= \int_G (\pi(gs)\xi|\pi(gs)\eta)dg = \int_G (\pi(g)\xi|\pi(g)\eta)dg = <\xi,\eta>.$$

en vertu de l'invariance à droite de la mesure dg.

Soient $\|.\|$ et $\|.\|_1$ les normes associées aux produits scalaires $(.|.)$ et $<.,.>$. D'après la continuité de π et la compacité de G, on a pour tout $\xi \in H_\pi$, $\sup_{g\in G}\|\pi(g)\xi\| < \infty$.

D'après le Théorème de Banach-Steinhaus, il existe une constante $C > o$ telle que pour tout $g \in G$ et pour tout $\xi \in H_\pi$, on ait

$$\|\pi(g)\xi\| \leq C\|\xi\|.$$

Si on applique cette inégalité au vecteur $\eta = \pi(g)^{-1}\xi$, il vient

$$\frac{1}{C}\|\xi\| \leq \|\pi(g)^{-1}\xi\|$$

pour tout $g \in G$ et pour tout $\xi \in H_\pi$.

Par intégration, on en déduit

$$\frac{1}{C}\|\xi\| \leq \|\xi\|_1 \leq C\|\xi\|$$

pour tout $\xi \in H_\pi$.

Donc les deux normes sont équivalentes. En particulier, H_π est complet pour la norme $\|.\|_1$ et π est continue pour cette norme.

Le résultat suivant va jouer un rôle important dans toute la suite de l'exposé.

<u>Théorème</u> 2.2.- <u>Soient</u> G <u>un groupe compact et</u> π <u>une représentation unitaire continue de</u> G <u>dans un espace hilbertien</u> H. <u>Si</u> v,v',w <u>et</u> w' <u>sont des éléments de</u> H, <u>posons</u>

$$\phi(v,w;v',w') = \int_G (\pi(g)v|w)\overline{(\pi(g)v'|w')}dg.$$

Alors

(i) il existe un opérateur hermitien borné $A : H \longrightarrow H$ tel que $A \circ \pi(g) = \pi(g) \circ A$ pour tout $g \in G$;

(ii) A est un opérateur compact.

Démonstration.- (i) : Pour w et w' fixés, $\phi(v,w;v',w')$ est une forme hermitienne continue sur $H \times H$. Il existe donc un opérateur borné $A : H \longrightarrow H$ tel que

$$\phi(v,w;v',w') = (A(v) | v').$$

Montrons que A est hermitien.

On a

$$(A(v)|v') = \int_G (\pi(g)v|w) \overline{(\pi(g)v'|w')} dg$$

$$= \overline{\int_G (\pi(g)v'|w') \overline{(\pi(g)v|w)} dg} = \overline{(A(v')|v)} = (v|A(v')).$$

D'où $A = A^*$.

L'invariance de la mesure dg par les translations à droite entraîne

$$\phi(\pi(s)v,w;\pi(s)v',w') = \phi(v,w;v',w')$$

c'est-à-dire

$$(A(\pi(s)v)|\pi(s)v') = (A(v)|v'),$$

d'où, puisque π est unitaire

$$A \circ \pi(s) = \pi(s) \circ A$$

pour tout $s \in G$.

(ii) : Montrons que A est un opérateur compact. Rappelons d'abord qu'un opérateur borné A d'un espace hilbertien H est dit compact s'il transforme une $(x_n)_{n \in \mathbb{N}}$ d'éléments de H, faiblement convergente en une suite $(Ax_n)_{n \in \mathbb{N}}$ fortement convergente, c'est-à-dire si la condition $\lim_{n \longrightarrow \infty} (x_n|y) = (x|y)$ pour tout $y \in H$ entraîne

$$\lim_{n \longrightarrow \infty} ||A(x_n) - A(x)|| = 0.$$

Comme

$$\lim_{n \to \infty} (Ax_n | Ax) = ||Ax||^2$$

et

$$||A x_n - A x||^2 = ||A x_n||^2 - (Ax_n | Ax) - (Ax | Ax_n) + ||Ax||^2,$$

il suffit de montrer que $\lim_{n \to \infty} ||A x_n|| = ||A x||$ pour voir que A est compact.

On a

$$\lim_{n \to \infty} ||A x_n||^2 = \lim_{n \to \infty} (Ax_n | Ax_n)$$

$$= \lim_{n \to \infty} \int_G (\pi^{-}(s) x_n | w) \overline{(\pi(s) Ax_n | w)} ds$$

$$= \lim_{n \to \infty} \int_G \int_G (\pi(s) x_n | w) (\pi(t) w | w) \overline{(\pi(ts) x_n | w)} dsdt.$$

D'après le Théorème de Banach-Steinhaus, une suite faiblement convergente est bornée. Il existe donc une constante $C > 0$ telle que $||x_n|| < \infty$ pour tout $n \in \mathbb{N}$. Alors, l'inégalité de Cauchy-Schwarz montre que

$$|(\pi(s) x_n | w) (\pi(t) w | w) \overline{(\pi(ts) x_n | w)}| \leq C^2.$$

Le théorème de convergence dominée de Lebesgue montre alors que

$$\lim_{n \to \infty} ||A x_n||^2 = \int_G \int_G (\pi(s) x | w) (\pi(t) w | w) \overline{(\pi(ts) x | w)} ds\, dt$$

$$= ||Ax||^2,$$

d'où la compacité de A.

Corollaire 2.3.- Toute représentation unitaire π d'un groupe compact G dans un espace hilbertien H est somme directe de représentations unitaires irréductibles de dimension finie. En particulier, toute représentation unitaire irréductible d'un groupe compact est de dimension finie.

Démonstration.- Soit H_λ le sous-espace propre de l'opérateur A associé à la valeur propre λ. D'après la théorie des opérateurs hermitiens compacts, H est somme directe des H_λ et $\dim(H_\lambda) < \infty$; de plus chaque H_λ est invariant par π puisque

$A \circ \pi(g) = \pi(g) \circ A$ pour tout $g \in G$.

Si on note π_λ la restriction de π à H_λ, on a donc

$$\pi = \underset{\lambda}{\oplus} \pi_\lambda .$$

Puisque chaque π_λ est une représentation unitaire de G de dimension finie, π_λ est somme directe de représentations unitaires irréductibles de dimension finie (on le voit facilement par récurrence sur la dimension de π_λ).

Si π est irréductible, l'opérateur A est scalaire. Donc dans H, l'opérateur identité est compact ce qui entraîne le fait que la boule unité de H qui est faiblement compacte est normiquement compacte. Donc H est de dimension finie.

Corollaire 2.4.- Soient G un groupe compact et π une représentation unitaire irréductible de G dans un espace hilbertien H de dimension finie n. Alors quels que soient les vecteurs v, v', w, w' de H, on a les relations de Schur :

$$\int_G (\pi(s)v|w)\overline{(\pi(s)v'|w')}\,ds = \frac{1}{n}(v|v')\overline{(w|w')} . \tag{2.1}$$

Démonstration.- L'opérateur A du Théorème 2.2 commute à la représentation π, donc d'après le lemme de Schur, il existe un scalaire λ tel que $A = \lambda I$, où I est l'application identique de H. On en déduit :

$$\phi(v,w;v',w') = \lambda(v|v') .$$

Mais on peut écrire

$$\begin{aligned} \phi(v,w;v',w') &= \int_G (v|\pi(s)^* w)\overline{(v'|\pi(s)^* w')}\,ds \\ &= \int_G \overline{(\pi(s)^* w|v)}(\pi(s)^* w'|v')\,ds \\ &= \overline{\int_G (\pi(s)w|v)\overline{(\pi(s)w'|v')}\,ds} \\ &= \overline{(A(w)|w')} = \bar{\lambda}\,\overline{(w|w')} \end{aligned}$$

D'où

$$\phi(v,w;v',w') = \lambda(v|v') = \overline{\lambda}\,\overline{(w|w')}.$$

Si on prend v = v' avec $\|v\| = 1$, on obtient

$$\lambda = \overline{\lambda}\,\overline{(w|w')}.$$

En prenant de même w = w' avec $\|w\| = 1$, on obtient

$$\overline{\lambda} = \lambda(v|v').$$

Donc

$$\phi(v,w;v',w') = \overline{\lambda}(v|v')\overline{(w|w')} = \lambda(v|v')\overline{(w|w')},$$ et par suite λ est un nombre réel positif car si v est un vecteur non nul,

$$\lambda\|v\|^4 = \lambda(v|v)\overline{(v|v)} = \phi(v,v;v,v) = \int_G |(\pi(s)v|v)|^2 ds \geq 0.$$

Montrons que $\lambda = \frac{1}{n} = \frac{1}{\dim(H)}$.

Soit $(e_1,\ldots,e_n)$ une base orthonormée de H. Alors $(\pi(s)e_1,\ldots,\pi(s)e_n)$ est encore une base orthonormée de H et on a l'égalité de Parseval

$$\sum_{i=1}^{n}(\pi(s)e_1|e_i)\overline{(\pi(s)e_1|e_i)} = (\pi(s)e_1|\pi(s)e_1) = (e_1|e_1) = 1.$$

En intégrant par rapport à s et utilisant l'égalité $\phi(v,w;v',w') = \lambda(v|v')\overline{(w|w')}$, il vient

$$\sum_{i=1}^{n}\int_G (\pi(s)e_1|e_i)\overline{(\pi(s)e_1|e_i)}ds = \int_G ds = 1$$

c'est-à-dire

$$\sum_{i=1}^{n}\lambda(e_1|e_1)(e_i|e_i) = n\lambda = 1.$$

D'où

$$\lambda = \frac{1}{n}.$$

Cas particulier .- Soit $A = (a_{ij}(s))_{i \leq i,j \leq n}$ la matrice de $\pi(s)$ par rapport à une base orthonormée $(e_1,\ldots,e_n)$ de H. On a

$$a_{ij} = (\pi(s)e_j | e_i),$$

d'où les <u>relations d'orthogonalité de Schur</u> :

(2.2) $$\int_G a_{ij}(s)\overline{a_{k\ell}(s)}\,ds = \int_G (\pi(s)e_j | e_i)\overline{(\pi(s)e_\ell | e_k)}\,ds = \frac{1}{n}\,\delta_{j\ell}\delta_{ik}.$$

§3.- <u>SERIE DE FOURIER SUR UN GROUPE COMPACT</u>

Soit G un groupe topologique compact. Si $\lambda \in \hat{G}$, soit π^λ un élément de λ de dimension $d(\lambda)$ et soit H l'espace de π^λ. Alors π^λ est représentée par une matrice unitaire

$$A = (a^\lambda_{ij}(s))_{1 \leq i,\, j \leq d(\lambda)}$$

par rapport à une base orthonormée $(e_1, \ldots, e_{d(\lambda)})$ de H.

On appelle <u>polynôme trigonométrique</u>, toute combinaison linéaire finie des $a^\lambda_{ij}(s)$ où $\lambda \in \hat{G}$ et $1 \leq i,j \leq d(\lambda)$.

Le résultat suivant généralise la notion classique de série de Fourier d'une fonction.

<u>Théorème</u> 3.1 (Peter-Weyl).- <u>Soit</u> G <u>un groupe compact</u>.

(i) <u>Pour tout</u> $\lambda \in \hat{G}$, <u>les fonctions</u>

(3.1) $$e^\lambda_{ij}(x) = \sqrt{d(\lambda)}\,a^\lambda_{ij}(x) \qquad 1 \leq i,j \leq d(\lambda)$$

<u>forment un système orthonormé total dans</u> $L^2(G)$.

(ii) <u>Toute fonction</u> $f \in L^2(G)$ <u>s'écrit</u>

(3.2) $$f(x) = \sum_{\lambda \in \hat{G}} d(\lambda) \sum_{i,j=1}^{d(\lambda)} (f | a^\lambda_{ij})\,a^\lambda_{ij}(x) \qquad (x \in G)$$

<u>où</u>

(3.3) $$(f | a^\lambda_{ij}) = \int_G f(x)\overline{a^\lambda_{ij}(x)}\,dx.$$

<u>La série</u> (3.2) <u>converge vers</u> f <u>dans</u> $L^2(G)$ <u>et on a la formule de Plancherel</u> :

$$(3.4) \qquad \int_G |f(x)|^2 dx = \sum_{\lambda \in \hat{G}} d(\lambda) \sum_{i,j=1}^{d(\lambda)} |(f|a_{ij}^{\lambda})|^2 .$$

Démonstration.- (i) : D'après les relations (2.2), les fonctions $e_{ij}^{\lambda}(x)$ forment un système orthonormé dans $L^2(G)$. Il reste donc à montrer que ce système est total, i.e. que l'ensemble des polynômes trigonométriques est dense dans $L^2(G)$.

Soit $f \in L^2(G)$ et soit $\varepsilon > 0$. D'après [1], Théorème 12.10, l'espace $\mathcal{C}(G)$ des fonctions continues sur G, muni de la norme de la convergence uniforme est dense dans $L^2(G)$. De même (cf.loc.cité, tome 2, Théorème 27.39) l'ensemble des polynômes trigonométriques est dense dans $\mathcal{C}(G)$.

Donc il existe $g \in \mathcal{C}(G)$ telle que

$$\|f - g\|_2 \leq \frac{\varepsilon}{2}$$

et il existe un polynôme trigonométrique h tel que

$$\|h - g\|_\infty \leq \frac{\varepsilon}{2} .$$

Alors

$$\|f - h\|_2 \leq \|f - g\|_2 + \|g - h\|_2 \leq \frac{\varepsilon}{2} + \frac{\varepsilon}{2} = \varepsilon.$$

(ii) : Soit $f \in L^2(G)$. Si on appelle coefficient de Fourier de f les nombres

$$C_{ij}^{\lambda} = \sqrt{d(\lambda)} \int_G f(x) \overline{a_{ij}^{\lambda}(x)}\, dx$$

et série de Fourier de f, la série de fonctions

$$\sum_{\lambda \in \hat{G}} \sum_{i,j=1}^{d(\lambda)} C_{ij}^{\lambda} e_{ij}^{\lambda}(x)$$

le reste du théorème est une conséquence immédiate de la théorie élémentaire des séries de Fourier dans un espace hilbertien.

On peut donner une forme invariante aux formules (3.2) et (3.4) en introduisant la notion de transformée de Fourier d'une fonction $f \in L^1(G)$.

Définition 3.2.- *Les notations étant celles qui ont été données au début de ce paragraphe, si* $f \in L^1(G)$, *on appelle transformée de Fourier de* f, *la fonction* $\hat{f}$ *définie sur* $\hat{G}$ *par la formule*

$$(3.5) \qquad \hat{f}(\lambda) = \int_G f(x)\pi^\lambda(x^{-1})dx$$

c'est-à-dire

$$(\hat{f}(\lambda)v|w) = \int_G f(x)(\pi^\lambda(x^{-1})v|w)dx$$

quels que soient $v, w \in H$.

Soient $A = (a^\lambda_{ij}(x))$ la matrice de π^λ par rapport à une base orthonormée $\left(e_1, \ldots, e_{d(\lambda)}\right)$ de H et $B = (b^\lambda_{ij})$ la matrice de l'opérateur $\hat{f}(\lambda)$ par rapport à la même base de H. On a

$$b^\lambda_{ji} = (\hat{f}(\lambda)e_i|e_j) = \int_G f(x)\overline{a^\lambda_{ij}(x)}dx = (f|a^\lambda_{ij}).$$

Alors

$$\mathrm{Tr}(\hat{f}(\lambda)\pi^\lambda(x)) = \sum_{i,k=1}^{d(\lambda)} b^\lambda_{ik}a^\lambda_{ki} = \sum_{i,k=1}^{d(\lambda)} (f|a^\lambda_{ki})a^\lambda_{ki}(x),$$

et la série de Fourier de f s'écrit :

$$f(x) = \sum_{\lambda\in\hat{G}} d(\lambda)\ \mathrm{Tr}(\hat{f}(\lambda)\pi^\lambda(x)).$$

Rappelons maintenant que la *norme de Hilbert-Schmidt* d'une matrice carrée complexe $A = (a_{ij})$ d'ordre n est définie par

$$||A||^2_{HS} = \sum_{i,j=1}^{n} |a_{ij}|^2 = \mathrm{Tr}(AA^*).$$

D'après ce qui précède, on a

$$||\hat{f}(\lambda)||^2_{HS} = \sum_{i,j=1}^{d(\lambda)} |(f|a^\lambda_{ij})|^2,$$

d'où la nouvelle forme de la formule de Plancherel

$$\int_G |f(x)|^2 dx = \sum_{\lambda \in \hat{G}} d(\lambda) \| \hat{f}(\lambda) \|_{HS}^2 .$$

Définition 3.3.- Soit π une représentation continue d'un groupe topologique G (non nécessairement compact) dans un espace vectoriel E de dimension finie n. On appelle caractère de π la fonction $\chi_\pi : G \longrightarrow \mathbb{C}$ définie par

$$(3.6) \qquad \chi_\pi(x) = \mathrm{Tr}\big(\pi(x)\big) \ .$$

Si $(e_1,\ldots,e_n)$ est une base de E et si

$$\pi(x)e_i = \sum_{j=1}^{n} a_{ji}(x)e_j ,$$

alors $$\chi_\pi(x) = \sum_{i=1}^{n} a_{ii}(x).$$

Le théorème suivant résulte aussitôt de la définition.

Théorème 3.4.- (i) Si π est une représentation unitaire de G, on a $\chi_\pi(x^{-1}) = \overline{\chi_\pi(x)}$ et $\chi_\pi(e) = \dim(\pi)$.

(ii) Deux représentations équivalentes ont même caractère.

(iii) Tout caractère χ est une fonction centrale sur G : $\chi(xy) = \chi(yx)$ quels que soient $x,y \in G$.

(iv) $\pi \oplus \pi'$ désignant la somme directe des représentations π et π', on a $\chi_{\pi\oplus\pi'} = \chi_\pi + \chi_{\pi'}$.

Théorème 3.5.- Soient π et π' deux représentations unitaires irréductibles d'un groupe compact G, χ et χ' les caractères de π et π' respectivement. Alors

(i) Si $\pi \not\simeq \pi'$, on a $\chi * \chi' = 0$.

En particulier

$$(\chi|\chi') = \int_G \chi(x)\overline{\chi'(x)}\,dx = 0.$$

(ii) <u>Si</u> $\pi \simeq \pi'$, <u>on a</u> $\quad \chi * \chi = \frac{1}{\dim(\pi)} \chi$.

<u>En particulier</u>

$$(\chi|\chi) = \int_G \chi(x)\overline{\chi(x)}\,dx = 1.$$

<u>Démonstration</u>.-(i) : Soit H (resp.H') l'espace hilbertien (de dimension finie) de la représentation π(resp.π'). Soit $(e_1,\ldots e_n)$ une base orthonormée de H et $(e'_1,\ldots,e'_m)$ une base orthonormée de H'. On a

$$\chi * \chi'(x) = \int_G \chi(y)\chi'(y^{-1}x)\,dy$$

$$= \sum_{i=1}^{n} \sum_{j=1}^{m} \int_G (\pi(y)e_i|e_i)(\pi'(y^{-1}x)e'_j|e'_j)\,dy$$

$$= \sum_{i=1}^{n} \sum_{j=1}^{m} \int_G (\pi(y)e_i|e_i)\overline{(\pi'(y)e'_j|\pi'(x)e'_j)}\,dy.$$

Donc si $\pi \not\simeq \pi'$, on a $\chi * \chi' = 0$ d'après les relations d'orthogonalité (il suffit de généraliser le Théorème 2.2)

$$\int_G (\pi(s)v|w)\overline{(\pi'(s)v'|w')}\,ds = 0$$

si $\pi \not\simeq \pi'$ où $v,w \in H$ et $v',w' \in H'$.

En particulier

$$(\chi|\chi') = \int_G \chi(y)\chi'(y)\,dy = \int_G \chi(y)\chi'(y^{-1})\,dx = \chi * \chi'(e) = 0.$$

(ii) : Si $\pi \simeq \pi'$, on peut prendre $\pi = \pi'$, $(e_i) = (e'_i)$ et on a

$$\chi * \chi(x) = \frac{1}{n} \sum_{i=1}^{n} \sum_{j=1}^{n} (e_i|e_j)\overline{(e_i|\pi(x)e_j)}$$

$$= \frac{1}{\dim(\pi)} \sum_{i=1}^{n} (\pi(x)e_i|e_i) = \frac{1}{\dim(\pi)} \chi(x).$$

En particulier, si $x = e$, on obtient

$$1 = \chi * \chi(e) = \int_G \chi(y)\overline{\chi(y)}\,dy = \|\chi\|_2^2.$$

Soit maintenant π une représentation de dimension finie d'un groupe compact G. Nous avons déjà observé que π était somme directe de représentations irréductibles π_i $(1 \leq i \leq k)$. On a

$$\chi_\pi = \chi_{\pi'} + \ldots + \chi_{\pi_k},$$

ce qui donne, en regroupant éventuellement les représentations équivalentes :

$$\chi_\pi = m_1 \chi_{\pi 1} + \ldots + m_\alpha \chi_{\pi_\alpha}.$$

m_i s'appelle la <u>multiplicité</u> de la représentation π_i dans π. On a d'après le Théorème 3.5,

$$m_i = (\chi_\pi | \chi_{\pi_i}).$$

<u>Exercices</u>

1) Soit G un groupe compact. Montrer que

(i) Une représentation unitaire π de dimension finie de G est irréductible si et seulement si $(\chi_\pi | \chi_\pi) = 1$.

(ii) Deux représentations unitaires π et π' de dimension finie de G sont équivalentes si et seulement si $\chi_\pi = \chi_{\pi'}$.

2) Soit G un groupe compact et soit $f \in L^2(G)$. Montrer que

(i) Si χ_λ est le caractère de π^λ, $\lambda \in \hat{G}$, la série de Fourier de f s'écrit en tout point $x \in G$:

$$f(x) = \sum_{\lambda \in \hat{G}} d(\lambda) f * \chi_\lambda (x) = \sum_{\lambda \in \hat{G}} d(\lambda) \chi_\lambda * f(x).$$

(ii) Si $g \in L^2(G)$, on a

$$(f|g) = \sum_{\lambda \in \hat{G}} d(\lambda) f * \tilde{g} * \chi_\lambda (e)$$

où $g(x) = \overline{g(x^{-1})}$.

BIBLIOGRAPHIE

[1] E. HEWITT et K. ROSS.- <u>Abstract harmonic analysis</u>, Springer New-York, tome 1, 1963 ; tome 2, 1970.

[2] M. NAIMARK et A. STERN.- <u>Théorie des représentations des groupes</u>, Editions Mir, 1979.

IV. Equations d'Einstein et d'Hamilton

LES EQUATIONS D'EINSTEIN

Francis CAGNAC

Département de Mathématiques, Universite de Yaoundé

Yaoundé, Cameroun.

I. INTRODUCTION

Les équations d'Einstein sont les équations fondamentales de la théorie de la relativité générale élaborée par Einstein en 1916.

Cette théorie rentre tout à fait dans le cadre de ce séminaire sur "les espaces fibrés et leur utilisation en physique", puisque l'outil essentiel de cette théorie est une variété différentiable et son fibré tangent, sans compter d'autres espaces fibrés ayant pour base la même variété.

Bien entendu, en 1916, l'expression "espace fibré" n'était pas encore en usage ; et la notion même de variété différentiable n'était pas encore clairement élaborée....

M'adressant à un public de mathématiciens, ignorant habituellement à peu près tout de la relativité générale, je me propose dans cet exposé :

1) d'introduire les équations d'Einstein- sans les "parachuter"- ce qui est le seul moyen de comprendre la théorie de la relativité générale.
2) de montrer, quels problèmes se posent à partir des équations d'Einstein et quel usage on peut en faire ;ceci, dans la mesure du temps qui me restera.

II. INTRODUCTION DES EQUATIONS D'EINSTEIN

Le point de départ d'Einstein fut la théorie de la relativité qu'il avait élaborée en 1905 et que nous appelons aujourd'hui "relativité restreinte" pour la distinguer de la théorie de la relativité générale.

La théorie de la relativité restreinte est la source de la relativité générale de deux manières :

- d'une part, parce que le point de départ de la réflexion d'Einstein est la représentation de l'univers donnée par la relativité restreinte.
- et d'autre part parce que c'est la même méthode de pensée qui avait conduit Einstein à élaborer la relativité restreinte, qui l'a ensuite conduit à élaborer la relativité générale.

En effet, l'origine de la relativité restreinte est la remarque suivante : les repères par rapport auxquels sont vérifiées les lois de la mécanique classique ne sont pas les mêmes que ceux par rapport auxquels sont vérifiées les lois de l'électromagnétisme.

L'échec de l'expérience de Michelson dans sa tentative de mettre en évidence le mouvement de la terre par rapport au repère dans lequel sont vérifiées les lois de l'électromagnétisme, a conduit Einstein à poser comme principe de base de la théorie de la relativité restreinte :

<u>les lois de la mécanique et les lois de l'électromagnétisme doivent être vérifiées par rapport aux mêmes repères.</u>

Le résultat est que

1) l'univers devient l'espace-temps, représenté par un espace de Minkowski E, c'est-à-dire un espace R^4 muni d'une forme quadratique g non dégénérée de signature + ---
2) il y a une correspondance bijective entre les repères galiléens usuels de la mécanique et les repères orthonormés de cet espace de Minkowski .

Mais cette représentation de l'univers fait encore jouer un rôle particulier à certains repères, les repères galiléens.

Et ceci n'est pas très satisfaisant pour l'esprit :

On prend habituellement un repère lié au soleil, mais pourquoi ne pas prendre un repère lié à n'importe quelle autre étoile ou galaxie ? Et toutes ces étoiles n'étant pas en mouvement de translation uniformes les unes par rapport aux autres, tous ne peuvent pas être des repères galiléens...

Ou encore, imaginez un univers où il n'y a qu'un seul astre. Suivant que cet astre est immobile ou en rotation par rapport à un repère galiléen les lois de la mécanique ne seront pas les mêmes sur cet astre. Mais s'il est seul dans l'univers, que signifie qu'il est en rotation ? Il tourne par rapport à quoi ?

De cette réflexion nait un des principes de base de la théorie de la relativité générale :

<u>les lois de la physique doivent pouvoir s'exprimer de la même maniere par rapport à n'importe quel repère</u>

Bien entendu, si on prend un nouveau repère lié à un corps qui n'est plus en translation uniforme par rapport au précédent repère, mais en mouvement quelconque les changements de coordonnées ne sont plus linéaires.

Cela conduit à poser que <u>les lois de la physique doivent s'exprimer de la meme manière par rapport à n'importe quel système de coordonnées curvilignes dans l'espace-temps</u>.

Ceci nous conduit à regarder comment s'expriment les lois de la mécanique de la relativité restreinte dans un système de coordonnées curvilignes.

En relativité restreinte, un point matériel est représenté dans l'espace de Minkowski par sa ligne d'univers L, lieu des évenements (x^α) qu'il occupe au cours du temps. Les vecteurs T tangents à L sont orientés dans le temps, c'est-à-dire, $g(T,T)>0$, et le vecteur unitaire tangent orienté vers les temps positifs, U, est appelé vecteur vitesse d'univers.

(Dans un repère galiléen, ses composantes U^α sont liés à la vitesse usuelle $v = (v^i)$

par $\begin{cases} U^0 = \dfrac{1}{\sqrt{1-v^2}} \\ U^i = \dfrac{v^i}{\sqrt{1-v^2}} \end{cases}$ en choisissant des unités telles que la vitesse de la lumière c=1)

- L'abscisse curviligne τ définie sur L par U s'appelle le temps propre du point matériel et la loi fondamentale de la mécanique du point matériel s'écrit:

(1) $$\frac{d}{d\tau}(m_0 U) = \Phi$$

m_0 = masse au repas du point matériel

Φ = vecteur force d'univers (dans un repère galiléen tel que $v^i = 0$, les composantes de Φ sont $\Phi^0 = 0$, $\Phi^i = f^i$, où f^i sont les composantes de la force au sens usuel).

Dans un système de coordonnées curvilignes x^λ, les grandeurs vectorielles ou tensorielles sont exprimées en chaque point par rapport au repère naturel (e_λ) associé à ces coordonnées. Et l'équation (1) devient :

(2) $$m_o \left(\frac{d^2x^\lambda}{d\tau^2} + \Gamma^\lambda_{\mu\nu} \frac{dx^\mu}{d\tau}\frac{dx^\nu}{d\tau}\right) = \phi^\lambda$$

où les $\Gamma^\lambda_{\mu\nu}$ sont les coefficients de la connexion euclidienne de E dans le système de coordonnées curvilignes x^λ.

$$\left(\text{ou } \frac{\partial e_\mu}{\partial x^\nu} = \Gamma^\lambda_{\mu\nu} e_\lambda \right)$$

(2) s'écrit aussi sous la forme :

(3) $$m_o \frac{d^2x^\lambda}{dt^2} = \phi^\lambda - m_o \Gamma^\lambda_{\mu\nu} \frac{dx^\mu}{d\tau}\frac{dx^\nu}{d\tau}$$

Sous cette forme, le terme supplémentaire qui apparait à droite, est ce qu'on appelle dans les cas usuels force d'entraînement, force de Coriolis, et d'une facon générale <u>forces d'inerties</u>.

Remarquons aussi que dans les cas usuels où

$$U^o = \frac{dx^o}{d\tau} \simeq 1 \text{ et } U^i = \frac{dx^i}{d\tau} \ll 1 \ (i=1,2,3)$$

ce terme supplémentaire est voisin de : $-m_o\Gamma^\lambda_{oo}$

Ceci nous amène au 2è principe de base de la relativité générale qui vient du problème de la <u>gravitation</u>.

Avec l'introduction de la relativité restreinte une reformulation de la loi de Newton, $f = k\frac{mm'}{r^2}$, était nécessaire car "r", la distance des 2 points matériels de masses m et m' n'est plus une grandeur invariante par changements de repères galiléens.

Mais la remarque d'Einstein au sujet de la gravitation est celle-ci:ce qu'il y a de remarquable dans la force de gravitation, c'est qu'elle est rigoureusement proportionnelle à la masse m du corps qui la subit.

Or c'est la propriété que possèdent les "forces d'inerties" qui, nous venons de le voir, s'introduisent quand on écrit les lois de la mécanique dans un autre système de coordonnées que les coordonnées cartésiennes définies par un repère galiléen.

D'où ce 2è principe de base de la théorie de la relativité générale :

<u>les forces de gravitation doivent pouvoir s'interpreter comme des forces d'inertie</u>

S'il en est ainsi, en l'absence de toute autre force, le mouvement d'un point matériel soumis à la seule gravitation est régi par les équations :

$$\frac{d^2x^\lambda}{d\tau^2} + \Gamma^\lambda_{\mu\nu} \frac{dx^\mu}{d\tau} \cdot \frac{dx^\nu}{d\tau} = 0,$$

c'est-à-dire les équations des géodésiques.

Bien entendu, si l'espace est euclidien, de telles géodésiques sont des droites de R^4. Or on sait qu'un point matériel soumis à la gravitation ne décrit pas une droite à vitesse constante : il suffit de penser au mouvement des astres ou au mouvement des projectiles.

On ne peut donc interpréter les forces de gravitation comme des forces d'inertie que si l'on admet que l'espace-temps n'est plus représenté par un espace euclidien, mais par une variété à 4 dimensions V^4, munie d'une pseudo-métrique g de signature +---, et de la connexion de Levi-Civita qui lui est associée.

Donc, en relativité générale, l'univers est représenté par une variété pseudo-riemanienne (V^4, g).
Et dans cette variété, la ligne d'univers d'un point matériel qui est soumis aux seules forces de gravitation est une géodésique de (V^4, g)

L'expérience et la loi de Newton montrent que la ligne d'univers d'un point matériel soumis aux seules forces de gravitation dépend de la répartition des masses dans l'univers :

$$\vec{F} = m\, k \sum \frac{m'_i}{r_i^2} \vec{u}_i$$

m → u_i → m_i

Il doit donc exister une loi liant d'une part les masses présentes dans l'univers, et d'autre part la pseudo-métrique g, c'est-à-dire, la géométrie de (V^4, g)

Pour trouver cette loi, Einstein fait appel à des raisons essentiellement mathématiques-

Puisqu'il s'agit d'une loi liant la métrique g et la matière présente dans l'univers, on pense à une équation de la forme :

(4) quelque chose qui dépend de g = quelque chose qui dépend de la matière.

Pour que cette loi s'exprime de la même manière par rapport à n'importe quel système de coordonnées, on pense évidemment à une égalité entre <u>grandeurs tensorielles</u>.

Pour ce qui est de la grandeur qui dépend de la matière, n'oublions pas qu'en relativité restreinte la masse n'est plus un invariant.

Les grandeurs qui caractérisent un état de la matière dans la mécanique de la relativité restreinte sont :

- soit le vecteur d'impulsion-énergie : $m_o U^\alpha$, pour un point matériel.

- soit le tenseur d'impulsion-énergie $T^{\alpha\beta}$, pour un milieu continu.

(Rappelons que $T^{\alpha\beta}$ est un tenseur symétrique d'ordre 2, et que, dans un repère galiléen, T^{oo} est la densité de masse, T^{oi} la densité d'impulsion, T^{ij} les tensions internes du milieu ; et que, dans les cas usuels, le terme prépondérant est T^{oo})

Et la loi fondamentale de la mécanique des milieux continus en relativité restreinte s'écrit :

(5) $$\partial_\alpha T^{\alpha\beta} = \phi^\beta$$

où ϕ^β est la densité de force d'univers extérieure s'exerçant sur le système.

Mais à partir du moment où les forces gravitation ne sont plus à considérer comme des forces extérieures et où l'expression de toutes les autres forces (tensions mécaniques ; électromagnétisme) peut rentrer dans le tenseur d'impulsion-énergie, il n'y a plus de forces extérieures.

Donc, en relativité générale, la matière est représentée par un champ de tenseurs symétriques d'ordre 2, $T^{\alpha\beta}$, appelé tenseur d'impulsion-énergie et astreint à vérifier l'égalité :

(6) $$\boxed{\nabla_\alpha T^{\alpha\beta} = 0}$$

((6) est l'égalité (5) où on a remplacé le 2è membre par 0, et la dérivée ordinaire, par une dérivée covariante)

Pour ce qui est de la grandeur qui dépend de g ;

Comme nous l'avons vu elle doit être tensorielle. Si l'on prend à droite dans (4) le tenseur d'énergie-impulsion $T^{\alpha\beta}$, il faut prendre à gauche un tenseur symétrique d'ordre 2 : Appelons le $S_{\alpha\beta}$; et l'équation (4) prend alors la forme :

(7) $S_{\alpha\beta} = \chi \ T_{\alpha\beta}$, χ constante.

Le tenseur $S_{\alpha\beta}$ doit être choisi de facon que, en 1e approximation, l'équation (7) redonne les lois de la gravitation newtonienne.

Or dans une portion de l'univers où - comme il est raisonnable de le penser pour le système solaire - on a un système de coordonnées tel que

$$g_{\lambda\mu} \approx \eta_{\lambda\mu}, \text{ métrique de Minkowski,}$$

pour des vitesses petites par rapport à la vitesse de la lumière dans ce système de coordonnées (c'est-à-dire $\frac{dx^0}{d\tau} \approx 1$, $\frac{dx^i}{d\tau} << 1$), l'équation des géodésiques s'écrit en 1e approximation, en posant $x^0 = t$, $\frac{d^2x^i}{dt^2} = -\Gamma^i_{00}$

- et si l'on suppose de plus que les $g_{\lambda\mu}$ ne dépendent pas du temps x^0, $-\Gamma^i_{00} \approx \partial_i \left(-\frac{1}{2} g_{00}\right)$

On retrouve l'équation d'un point matériel dans un champ de gravitation si $-\frac{1}{2} g_{00} = U$, potentiel newtonien.

Or U est déterminé par l'équation de Poisson $\Delta U = -4\pi k\rho$,

où ρ est la densité de masse, égale en 1e approximation à la composante T_{00} du tenseur d'impulsion-énergie.

On trouve donc que (7) redonnera en approximation les lois de la gravitation newtonienne, si, en 1e approximation, (7) entraîne :

(8) $\Delta \left(\frac{1}{2} g_{00}\right) = 4\pi k \ T_{00}$

Pour avoir des chances que (7) entraîne (8), il faut que les

(C) composantes du tenseur $S_{\alpha\beta}$ s'expriment au moyen des $g_{\lambda\mu}$ et de leurs dérivées 1ères et secondes et soient linéaires par rapport aux dérivées secondes

On pense alors au tenseur de courbure $R^\lambda{}_{\mu,\nu\ell}$

et à ses contractés :

le tenseur de Ricci $R_{\alpha\beta} = R^\lambda{}_{\alpha,\lambda\beta}$

et la courbure scalaire $R = R^\alpha_\alpha$

Elie Cartan a d'ailleurs montré qu'il n'y a pas d'autres tenseurs d'ordre 2 vérifiant les conditions ci-dessus que les combinaisons linéaires à coefficients constants de $R_{\alpha\beta}$, $Rg_{\alpha\beta}$, et $g_{\alpha\beta}$

La première idée d'Einstein fut de choisir le tenseur de Ricci et de poser ;

$$R_{\alpha\beta} = X\ T_{\alpha\beta}$$

Mais nous avons vu que le tenseur d'impulsion-énergie $T_{\alpha\beta}$ doit vérifier la condition $\nabla_\alpha T^{\alpha\beta} = 0$.

Le tenseur figurant à gauche dans (7) doit évidemment vérifier la même condition, c'est-à-dire.

(9) $$\nabla_\alpha S^{\alpha\beta} = 0$$

Ce n'est pas le cas du tenseur de Ricci, mais Einstein a trouvé les tenseurs vérifiant simultanément (C) et (9). Ce sont

(10) $$\boxed{S_{\alpha\beta} - R_{\alpha\beta} - \frac{1}{2} g_{\alpha\beta} R + \Lambda\, g_{\alpha\beta}},\quad \Lambda \text{ constante.}$$

On démontre aisément au moyen des identités de Bianchi que les tenseurs $S_{\alpha\beta}$ définis par (10) vérifient (9) et Elie Cartan et Hermann Weyl ont d'ailleurs montré que ce sont les seuls tenseurs vérifiant simultanément les conditions (C) et (9).

Les équations d'Einstein s'écrivent donc :

(11) $$\boxed{S_{\alpha\beta} = X\ T_{\alpha\beta}}$$

La constante X est liée à la constante k de l'attraction universelle de Newton, par la condition que, en 1e approximation, (11) doit entraîner (8).

Or en faisant les mêmes approximations que ci-dessus, et en prenant en outre en 1e approximation, $T_{oo} = \rho$, $T_{oi} = T_{ij} = 0$, on trouve que l'équation $S_{oo} = X\ T_{oo}$, devient :

$$\Delta\left(\frac{1}{2} g_{oo}\right) = \frac{1}{2} X\ T_{oo} + \Lambda$$

On retrouve (8) à condition que

1) la constante Λ soit petite, négligeable en 1e approximation

2) $X = 8\pi k$ (ou si $c \neq 1$, $X = \frac{8\pi k}{c^4}$).

Résumons nous:

Le tenseur $S_{\alpha\beta}$, défini par (10) est appelé <u>tenseur d'Einstein</u>

La constante Λ qui y figure est appelée <u>constante cosmologique</u>.

Elle est habituellement supposée nulle et c'est ce que nous ferons.

<u>En relativité générale, le tenseur d'impulsion-énergie</u> et le <u>TENSEUR d' Einstein sont liés par les équations d'Einstein</u>

$$S_{\alpha\beta} = X\ T_{\alpha\beta}$$

III. UTILISATION DES EQUATIONS D'EINSTEIN

Comment ces équations permettent-elles de décrire l'univers ?

Nous avons vu que la matière présente dans l'univers est représentée dans les équations d'Einstein par le tenseur d'impulsion-énergie.

Pour trouver celui-ci, la corespondance avec les données usuelles en relativité restreinte est la suivante :

> Au voisinage d'un point $x \in V^4$, les repères orthonormées de $T_x V$ correspondent aux repères galiléens de la relativité restreinte; toute l'interprétation physique de la relativité générale repose sur ce principe.

Donnons en quelques exemples usuels :

En mécanique des fluides, l'état de la matière est représenté en relativité restreinte par une fonction $\rho(x^\lambda)$: densité de masse au repos en x.
et par un champ de vecteurs : $u^\alpha(x^\lambda)$: vitesse d'univers du point matériel situé en x .

- Si on néglige les tensions, il lui correspond un tenseur d'impulsion-énergie:

$$T^{\alpha\beta} = \rho\, u^\alpha u^\beta$$

- S'il y a des tensions, le tenseur d'impulsion-énergie est :

$$T^{\alpha\beta} = \ell\, u^\alpha u^\beta + \Theta^{\alpha\beta}$$

($\Theta^{\alpha\beta}$ tenseur qui depend des tensions; dans un repère tel que au point x^α, $u^0 = 1$, $u^j = 0$, on a $\Theta^{ij} = T^{ij}$, $\Theta^{0\alpha} = 0$).
En particulier si les tensions se réduisent à une pression p, (fluide parfait),

$$T^{\alpha\beta} = (\rho+p)u^\alpha u^\beta - p\eta^{\alpha\beta}$$

= En relativité générale on représentera donc aussi une telle matière par une fonction ρ et un champ de vecteurs unitaires orientés dans le temps u^α. Avec un tenseur d'énergie-impulsion $T^{\alpha\beta} = \rho u^\alpha u^\beta$ si on néglige les tensions,

$$\text{et} \quad T^{\alpha\beta} = (\rho+p)u^\alpha u^\beta - pg^{\alpha\beta}, \text{ pour un fluide parfait}$$

Dans toute carte locale, cela fait un système d'au moins 10 équations aux dérivées partielles du second ordre (les équations d'Einstein) avec au moins 10 fonctions inconnues, les $g_{\lambda\mu}$,...

Problèmes locaux et problèmes globaux

On peut soit chercher à faire un modèle de tout l'univers, soit un modèle d'une partie de l'univers.

Faire un modèle de tout l'univers, c'est ce que l'on fait dans les théories cosmologiques ; théories de l'univers en expansion, "big bang" initial, etc... ; c'est là que se pose le problème de la nature topologique de la variété V.

Faire un modèle d'une partie de l'univers; un cas typique est faire un modèle du système solaire. Dans ce cas on prend une variété V homéomorphe à $\mathbb{R}^4$ et on admet que dans l'infini spatial l'univers est plat, ce qui revient à admettre qu'il existe des systèmes de coordonnées pour lesquels $g_{\alpha\beta}(x) \longrightarrow \eta_{\alpha\beta}$ quand x tend vers l'infini dans une direction spatiale - Mathématiquement, c'est encore un problème global -

Mais on peut aussi se poser des problèmes locaux : le cas typique est celui du problème de Cauchy.

Etant donné une hypersurface S de V^4 qui partage V^4 en 2 régions I et II.

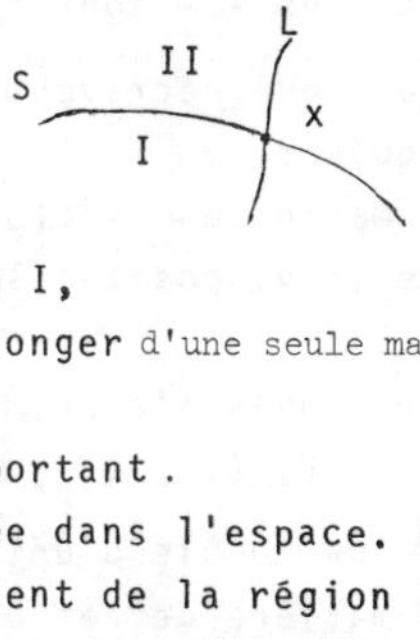

Etant donné une solution du système des équations d'Einstein et des équations physiques dans la région I, peut-on la prolonger - et peut-on la prolonger d'une seule manière dans la région II ?

Physiquement c'est un problème important.

Supposons l'hypersurface S orientée dans l'espace. Cela signifie que toutes les lignes de temps L passent de la région I à la région II en traversant S ; la région II peut donc être considérée comme "postérieure" à I. Et le problème posé revient à celui-ci : la partie II de l'univers est-elle déterminée par la partie I ?

- Un champ électromagnétique est décrit en relativité restreinte par un champ de 2 tenseurs antisymétriques $F_{\lambda\mu}$ vérifiant :

$$\partial_\alpha F_{\beta\gamma} + \partial_\beta F_{\gamma\alpha} + \partial_\gamma F_{\alpha\beta} = 0$$ (1e groupe d'équations de Maxwell)

et donnant lieu à un tenseur d'impulsion énergie

$$T^{\alpha\beta} = -F^{\lambda\alpha}F^{\beta}_{\lambda} + \frac{1}{4}\eta^{\alpha\beta}F^{\lambda\mu}F_{\lambda\mu}\ .$$

En relativité générale ce champ sera représenté par un champ de 2 tenseurs antisymétriques $F_{\lambda\mu}$ vérifiant :

$$\nabla_\alpha F_{\beta\gamma} + \nabla_\beta F_{\gamma\alpha} + \nabla_\gamma F_{\alpha\beta} = 0$$

et il lui correspondra un tenseur d'impulsion-énergie

$$T^{\alpha\beta} = -F^{\lambda\alpha}F^{\beta}_{\lambda} + \frac{1}{4}g^{\alpha\beta}F^{\lambda\mu}F_{\lambda\mu}$$

Un modele d'univers consiste donc en une variété (V^4, g) dans laquelle la matière et le champ électromagnétique sont décrits par des fonctions ou des champs.

ρ, u^α, p, $F_{\lambda\mu}$, etc... ; appelons les v_r

suivant les parties de l'univers on peut d'ailleurs avoir des qualités de matières différentes ; on peut avoir le vide dans certaines parties.

A cette matière correspond un champ de tenseur d'impulsion énergie : $T^{\alpha\beta}$ où les $T^{\alpha\beta}$ sont fonctions des v_r et des $g_{\lambda\mu}$: $T^{\alpha\beta}(v_r, g_{\lambda\mu})$.

Les fonctions qui décrivent cette matière peuvent être liées par des lois physiques :

loi de Mariotte $\rho = \phi(p)$

lois de la viscosité, liant les tensions et le champ des vitesses etc...

Ces lois physiques s'exprimeront par les équations :

$$F_k(v_r, g_{\lambda\mu}) = 0.$$

On aura un modèle d'univers si les fonctions et champs v_r qui décrivent la matière de cet univers et le tenseur métrique $g_{\lambda\mu}$, vérifient :

$$(12)\quad \begin{cases} S_{\alpha\beta}(g_{\lambda\mu}) = XT_{\alpha\beta}(v_r, g_{\lambda\mu}) & \text{les équations d'Einstein} \\ F_k(v_r, g_{\lambda\mu}) = 0 & \text{et les équations de la physique} \end{cases}$$

IV. CONCLUSIONS

Ce qui a fait le succès de la théorie de la relativité générale d'Einstein, outre les quelques vérifications expérimentales qu'elle a rencontrées, c'est que ses équations "s'imposent", sans aucun arbitraire, dès qu'on a admis les 2 principes très généraux qui sont à la base de cette théorie, à savoir l'invariance par tout changement de coordonnées curvilignes et le principe des géodésiques parcourues par un point matériel soumis à la seule gravitation.

Je termine en donnant une liste de quelques livres qui permettent à des mathématiciens de s'initier à cette théorie ; ils sont tous en anglais, car en francais, il n'en existe pas d'autre à ma connaissance, que le vieux livre de A. Lichnérowicz "Théories relativistes de la gravitation et de l'électromagnétique" (Masson, 1955)

REFERENCES

S. Hawking et G. Ellis, The Large Scale Structure of Space-Time, (Cambridge, 1979).

Achille Papapetrou, Lectures on General Relativity (Reidel, 1974).

Sachset Wu, General Relativity for Mathematicians (Springer 1977)

Bernard F. Schutz, A First Course in General Relativity (Cambridge 1985).

SYSTEMES HAMILTONIENS

Cherif BADJI

Departement de Mathématiques, Université de Dakar, Senegal.

I. INTRODUCTION

Les succès du formalisme hamiltonien en mécanique classique (description complète de la mécanique classique en termes de géométrie symplectique et de parenthèse de Poisson) ont motivé les tentatives (avec parfois beaucoup de succès) de son extension à diverses branches de la physique : mécanique quantique, théorie classique des champs (par exemples : équation des ondes, linéaire et non linéaire, équation de MAXWELL), hydrodynamique des fluides parfaits, relativité générale etc. Ceci et les résultats profonds de EGOROV, HÖRMANDER, NIRENBERG et TREVES concernant les équations aux dérivées partielles linéaires (grâce auxquels la géométrie symplectique et la mécanique classique ont apporté un souffle nouveau aux E.D.P.) expliquent l'intérêt croissant manifesté ces dernières décades par de nombreux mathématiciens (surtout mathématiciens-physiciens) à l'endroit de la géométrie symplectique, des systèmes hamiltoniens et des algèbres de Lie classiques de dimension infinie.

Nous nous contenterons, ici, d'un survol de résultats classiques ainsi que de méthodes et résultats de ces dix dernières années dûs à LICHNEROWICZ, AVEZ et DIAZ-MIRANDA dans le cas des systèmes hamiltoniens classiques et à J. MARSDEN et P. CHERNOFF dans le cas des systèmes hamiltoniens linéaires et non linéaires (résultats et méthodes non exhaustifs cependant).

Le lecteur désireux d'approfondir ses connaissances sur le programme de quantisation de A. LICHNEROWICZ (déformations d'algèbre de Lie, variétés de Jacobi, variétés de Poisson, * - Produit) peut se reférer à [1], et les références qui y sont.

II. VARIETES SYMPLECTIQUES

II-1 - Espaces vectoriels symplectiques

Soit E un $\mathbb{R}$-espace vectoriel de dimension n. Soit α une 2-forme extérieure sur E. De considérations classiques, il existe un entier pair $2s \leq n$ (cf. [5]) et $2s$ formes linéaires indépendantes α_i, $1 \leq i \leq 2s$, sur E telles que :

(1.1) $\alpha = \alpha_1 \wedge \alpha_2 + \ldots + \alpha_{2s-1} \wedge \alpha_{2s}$.

Ainsi α est de rang maximum $2s$ si et seulement si $\alpha^s \neq 0$ et $\alpha^{s+1} = 0$ (cf. [5]).

1.2 - Définition : Une structure symplectique sur E est définie par la donnée d'une 2-forme extérieure α de rang maximum n sur E (ici nécessairement E est de dimension paire $n = 2m$ car le rang d'une 2-forme extérieure est paire d'après ce qui précède).

Le couple (E, α) est appelé espace vectoriel symplectique. Les propriétés suivantes sont équivalentes :

(i) (E, α) est un espace vectoriel symplectique

(1.3) (ii) α^m est une forme volume sur E

(iii) L'application $x \longrightarrow i_x(\alpha)$ de E sur E^*, où $i(\ )$ désigne le produit intérieur, est un isomorphisme d'espaces vectoriels.

1.4 - Définition : Soient (E, α) et (F, β) deux $\mathbb{R}$-espaces vectoriels de même dimension $n = 2m$. Un isomorphisme symplectique

$$h : (E, \alpha) \xrightarrow{\sim} (F, \beta)$$

est un isomorphisme linéaire $h : E \xrightarrow{\sim} F$ tel que $h^* \beta = \alpha$.

Soit (E, α) un espace vectoriel symplectique. L'ensemble $S_p(E, \alpha)$ des automorphismes symplectiques de (E, α) est un sous-groupe du groupe $SGL(E)$ des automorphismes de déterminant égal à 1 de E.

Soit maintenant $\mathcal{B}$ un espace de Banach (réel) et $\Omega : \mathcal{B} \times \mathcal{B} \longrightarrow \mathbb{R}$ une forme bilinéaire continue sur $\mathcal{B}$. Pour tout point $e \in \mathcal{B}$ l'application $\Omega_e : \mathcal{B} \longrightarrow \mathcal{B}^*$ définie par :

$$\forall f \in \mathcal{B} \longrightarrow \Omega_e(f) = \Omega(e, f)$$

est linéaire et continue.

1.5 - Définition : La forme bilinéaire $\Omega : \mathcal{B} \times \mathcal{B} \longrightarrow \mathbb{R}$ est dite faiblement non dégénérée si $\forall e \in \mathcal{B}$ l'application Ω_e est injective (i.e. : $\Omega(e, t) = 0$ $\forall f \in E \Longrightarrow e = 0$).

1.6 - Remarque : La non-dégénérécence faible et forte sont confondues dans le cas où $\mathcal{B}$ est de dimension finie. La distinction est cependant importante dans le cas de la dimension infinie.

II-2 - Structures symplectiques sur une variété

A) - Nous supposerons d'emblée dans ce paragraphe que tous les éléments qui interviennent (variétés, fonctions, formes différentielles et champs de vecteurs sur les variétés) sont de classe C^∞.

2.1 - <u>Définition</u> : Soit M une variété différentielle réelle de dimension $2n$. Une structure symplectique sur M est définie par la donnée d'une 2-forme différentielle fermée F ($dF = 0$) partout de rang $2n$ sur M.

(2.2) - <u>Exemple</u> : Soit le $\mathbb{R}$-espace vectoriel $\mathbb{R}^{2n}$ muni des coordonnées p_i, q_i, $1 \leqslant i \leqslant n$. On vérifie immédiatement que la 2-forme différentielle

$$(2.3) \qquad F = \sum_{i=1}^{n} d\,p_i \wedge d\,q_i \ ,$$

est fermée et de rang constant $2n$. Ainsi $(\mathbb{R}^{2n}, F)$ est une variété symplectique.

(2.4) - <u>Exemple</u> : Soit M une variété différentielle C^∞ de dimension n (non nécessairement paire), $T(M)$ et $T^*(M)$ les fibrés tangent et cotangent (respectivement) de M ; $T_x(M)$ et $T^*_x(M)$ désigneront les fibres au-dessus de $x \in M$. Soit $\Pi : T^*(M) \longrightarrow M$ la projection de $T^*(M)$ sur la base M. Alors $T\Pi : T(T^*M) \longrightarrow TM$ est la dérivée de l'application Π. Soit $u \in T_\alpha(T^*(M))$ un vecteur tangent à $T^*(M)$ au point $\alpha \in T^*_x(M)$. Moyennant la dérivée $T\Pi : T(T^*(M)) \longrightarrow T(M)$ on définit une forme linéaire sur $T_\alpha(T^*(M))$ par :

$$u \longrightarrow \langle T\Pi(u), \alpha\rangle = \langle u, (T\Pi)^*(\alpha)\rangle ,$$ où $(T\Pi)^*$ désigne la transposée de $T\Pi$. La correspondance $\alpha \longrightarrow (T\Pi)^*(\alpha)$ définit donc une forme de Pfaff λ de classe constante $2n$ sur $T^*(M)$. La 1-forme $\lambda = (T\Pi)^*(\alpha)$ est appelée 1-forme de Liouville sur $T^*(M)$. Si $q = (q_1, \ldots, q_n)$ est une collection de coordonnées locales dans un ouvert de coordonnées U en x alors un point $\alpha \in T^*(M)$ est déterminée par ses n composantes $p = (p_1, \ldots, p_n)$. Le couple (p, q) forme une collection de coordonnées locales de $\alpha \in T^*M$. Dans ces coordonnées locales (p, q) la forme λ s'écrit :

$$\lambda = p\,d\,q = \sum_{i=1}^{n} p_i\,d\,q_i$$

(2.5) d'où

$$d\lambda = d\,p \wedge d\,q = \sum_{i=1}^{n} d\,p_i \wedge d\,q_i \ .$$

On vérifie aisément que $d\lambda$ est fermée et de rang constant 2n.

Ainsi $(T^*(M), d\lambda)$ est une variété symplectique.

Soit (M, F) une variété symplectique de dimension $2n$. Il résulte de considérations classiques que $F^n \neq 0$ (partout sur M) et donc définit un élément de volume $\eta = \frac{F^n}{n!}$ sur M (l'élément de volume symplectique). Une variété symplectique de dimension finie est donc orientable. On sait (Théorème de DARBOUX cf [4], [5]) que M admet des atlas de cartes locales dites canoniques (p, q), où $p = (p_1, \ldots, p_n)$ et $q = (q_1, \ldots, q_n)$ telles que F puisse s'écrire sur le domaine d'une de ces cartes :

$$(2.6) \qquad F|_{\mathcal{V}} = dp \wedge dq = \sum_{i=1}^{n} dp_i \wedge dq_i .$$

Une structure symplectique est, au sens des G-structures une $SP(n, \mathbb{R})$ - structure intégrable ; $SP(n, \mathbb{R})$ désigne le groupe symplectique réel à 2n variables. Notons que pour tout point $x \in M$, $(T_x(M), F(x))$ est un espace vectoriel symplectique.

L'isomorphisme $\mu_x : T_x(M) \xrightarrow{\sim} T_x^*(M)$, défini $\forall X \in T_x(M)$ par $\mu_x(X) = i_X(F(x))$, détermine un isomorphisme différentiable $\mu : T(M) \xrightarrow{\sim} T^*(M)$ du fibré tangent $T(M)$ sur le fibré cotangent $T^*(M)$. En particulier, si $\chi(M)$ est l'algèbre de Lie des champs de vecteurs différentiables sur M et $\Lambda^1(M)$ l'espace des 1-formes différentielles différentiables sur M, μ_x induit un isomorphisme (d'espaces vectoriels) de $\chi(M)$ sur $\Lambda^1(M)$
$(\forall X \in \chi(M) \longrightarrow \mu(X) = i_X(F) \in \Lambda^1(M))$. De plus l'isomorphisme $\mu : T(M) \xrightarrow{\sim} T^*(M)$ s'étend naturellement aux fibrés des tenseurs ; en ce qui concerne les tenseurs anti-symétriques, il est compatible avec le produit extérieur. Pour deux p-formes α et β on a :

$$(2.7) \qquad i(\mu^{-1}(\alpha))\, \beta = (-1)^p\, i(\mu^{-1}(\beta))\, \alpha$$

Soit $\mathcal{N} = C^\infty(M)$ l'algèbre associative réelle des fonctions C^∞ réelles sur M ; $\mathcal{N}_0$ désignera la sous-algèbre de $\mathcal{N}$ qui consiste en les fonctions à support compact. Enfin $\mathcal{N}_1$ désignera le sous-espace de $\mathcal{N}_0$ qui consiste en les fonctions $\mu \in \mathcal{N}_0$ telles que

(2.8) $$\int_M \mu\eta = 0 .$$

On sait que si M est compacte, l'élément de volume η ne peut être une forme exacte ; il s'en suit que la 2-forme différentielle F (et toutes ses puissances) ne peut être une forme exacte.

Soit $*$ l'opérateur d'adjonction symplectique défini pour toute p-forme α par :

(2.9) $$*\alpha = \mathcal{i}(\mu^{-1}(\alpha))\eta$$

Si α et β sont deux formes arbitraires on a

(2.10) $$(\alpha \wedge *\beta) = *\mathcal{i}(\mu^{-1}(\beta))(\alpha)$$

On sait que $*^2 = Id$ [2] . Si L est l'opérateur de degré 2 sur les formes défini par le produit extérieur par F $(L\alpha = F \wedge \alpha)$ on a l'opérateur Λ de degré -2 défini par :

(2.11) $$\Lambda = *^{-1} L *$$

D'après (2.10) on a : $\Lambda\alpha = *(F \wedge \alpha) = \mathcal{i}\mu^{-1}(F)(*^2\alpha)$, c'est-à-dire :

(2.12) $$\Lambda = \mathcal{i}\mu^{-1}(F)$$

De l'opérateur d de différentiation extérieure sur les formes on déduit l'opérateur δ de codifférentiation symplectique, défini pour toute p-forme α par :

(2.13) $$\delta\alpha = (-1)^p *^{-1} d * \alpha$$

On a immédiatement [2] :

(2.14) (a) $\delta^2 = 0$; (b) $d\Lambda - \Lambda d = \delta$; (c) $\delta L - L\delta = -d$

B) - Supposons maintenant que M est paracompacte connexe. D'après Lichnérowicz [2] il existe des métriques g échangeables avec F, i.e. telles que $\mu^{-1}(g)$ soit le tenseur contravariant inverse de g. Si $\nu : X \longrightarrow \mathcal{i}_X(g)$ est la dualité définie par la métrique g alors $\mu^{-1} \circ \nu$ détermine sur M un opérateur $\mathcal{J}$ de structure presque complexe. On obtient ainsi sur (M, F) une structure (M, F, g) dite presque kählérienne subordonnée à la structure symplectique envisagée. Soit $\bar{\delta}$ l'opérateur de codifférentiation métrique. Alors d'après [2] on a $\bar{\delta}F = 0$. Ainsi sur une variété munie d'une structure presque kählérienne la 2-forme symplectique F est à la fois fermée et cofermée $(dF = 0 = \bar{\delta}F)$.

Pour une 1-forme α la formule (2.14 ; (b)) se réduit à :

(2.15) $$\Lambda d\alpha = \delta\alpha ;$$

soit, F étant cofermée,

(2.16) $$\Lambda d\alpha = \nabla_i (F^{ij} \alpha_j)$$

où ∇ est l'opérateur de dérivation covariante dans la connexion riemannienne définie par la métrique g . D'après (2.16) l'opérateur δ est une divergence métrique. Ainsi pour une 1-forme α à support compact on a [2], [12]:

(2.17) $$\int_M \Lambda \ d\alpha = 0$$

III - SYSTEMES HAMILTONIENS CLASSIQUES

III.1 - Soit (M, F) une variété symplectique C^∞ de dimension 2n. Etudions l'action de l'algèbre $\chi(M)$ sur la 2-forme fondamentale F . Désignons par $\mathcal{L}_X$ l'opérateur de dérivation de Lie. Alors :

(1.1) $$\mathcal{L}_X F = d\, i_X F + i_{(X)}\, dF ;$$

soit, F étant fermée :

(1.2) $$\mathcal{L}_X F = d\, i_X F .$$

$X \in \chi(M)$ définit une transformation infinitésimale symplectique (ou est un champ de vecteurs hamiltonien) si F reste invariante par X . Pour qu'il en soit ainsi il faut et il suffit, d'après (1.2) ci-dessus, que $\mu(X) = i_X F$ soit fermée.

<u>Définition</u> : Un système hamiltonien sur une variété symplectique (M, F) est un champ de vecteurs $X \in \chi(M)$ tel que $i_X(F)$ soit une 1-forme fermée.

Si $i_X(F)$ est exacte, un hamiltonien H de X est une fonction $H \in \mathcal{N} = C^\infty(M)$ telle que $i_X F = dH$. Si M est connexe deux hamiltoniens de X diffèrent donc d'une constante. Réciproquement si α (respectivement H) est une 1-forme fermée (respect. un élément de $\mathcal{N}$) sur M, il existe un système hamiltonien X sur M et un seul tel que $i_X F = \alpha$ (respect. $i_X F = dH$). X est alors dit système hamiltonien associé à α (respect. à H). Si $i_X F = \alpha$ on a :

(1.3) $$i_X F(X) = d(X) = F(X, X) = 0$$

Ainsi la 1-forme α est une intégrale première de X. En particulier pour $i_X F = dH$ on a :

(1.4) $$i_X F(X) = dH(X) = X(H) = F(X, X) = 0 ;$$

donc H est intégrale première de X. H est appelé l'intégrale de l'énergie.

1.5 - <u>Définition</u> : Une application $g : M \longrightarrow M$ est dite symplectique si

$$g^* F = F ;$$

i.e, la 2-forme F est invariant intégral de g.

Si $M = \mathbb{R}^{2n}$, une application $g : \mathbb{R}^{2n} \longrightarrow \mathbb{R}^{2n}$ qui conserve la forme canonique $F = dp \wedge dq$ (voir exemple 2.2 de II) est dite application canonique de $(\mathbb{R}^{2n}, F)$.

Pour $H \in \mathcal{N}$ nous noterons X_H le champ de vecteurs $\mu^{-1}(dH)$ correspondant à dH. Supposons que X_H engendre un groupe à un paramètre $\phi_t : M \xrightarrow{\sim} M$ de difféomorphismes de M. Alors on a :

$$(1.6) \qquad \frac{d}{dt} \phi_t(x) \Big|_{t=0} = X_H \ ;$$

le flot $\{\phi_t\}_{t \in \mathbb{R}}$ est appelé flot hamiltonien de fonction de Hamilton H. Nous avons dans [4] le :

1.7 - Théorème : Le flot hamiltonien $\{\phi_t\}_{t \in \mathbb{R}}$ conserve la structure symplectique :

$$\phi_t^*(F) = F$$

De (1.4) découle le théorème de conservation de l'énergie suivant :

1.8 - Théorème : Le flot $\{\phi_t\}_{t \in \mathbb{R}}$ de hamiltonien H admet H pour intégrale première.

Soit L l'ensemble des champs de vecteurs hamiltoniens (ou t.i. symplectiques). Muni du crochet de Lie, L est une sous-algèbre de Lie de $\chi(M)$. Nous noterons L_o la sous-algèbre de Lie de L qui consiste en les champs de vecteurs hamiltoniens $X \in L$ à support compact, L^* le sous-espace de L défini par les champs de vecteurs dont l'image par μ est une 1-forme exacte. Un élément de L^* est encore dit champ de vecteurs globalement hamiltonien. Enfin nous désignerons par L_o^* (resp. L_1) le sous-espace de L_o défini par les vecteurs X tels que $\mu(X) = dg$ avec $g \in \mathcal{N}_o$ (resp. $\mu(X) = du$ avec $u \in \mathcal{N}_1$). Pour tout X, $Y \in L$ on a $L(X)F = L_Y F = 0$ et il vient par un calcul facile ([3]) :

$$(1.9) \qquad \mu[X, Y] = d \wedge \mu(X) \wedge d\mu(Y)$$

Définition : Un champ de vecteurs $X \in \chi(M)$ est appelé transformation infinitésimale conforme symplectique s'il existe $a \in \mathcal{N}$ telle que :

$$\mathcal{L}_X F + aF = 0 \ ;$$

c'est-à-dire : $d\, \imath_X F + aF = 0$. D'où par différentiation :

(1.10) $da \wedge F = 0$. F étant de rang maximum, d'après un théorème de Lepage, pour $n > 1$, (1.10) entraîne $da = 0$ c'est-à-dire si M est connexe, que $a = \text{constante} = K_X$ (a est localement constant si M n'est pas connexe). En accord avec LICHNEROWICZ [3] en supposant désormais que M est connexe, nous noterons L^c et appelerons (par abus de langage) algèbre de Lie des t.i. conformes symplectiques l'algèbre de Lie des champs de vecteurs X tels que

$\mathcal{L}_X F + K_X F = 0$, où K_X est une constante qui dépend de X .

Pour tout $X \in L^c$ et tout $Y \in L$ la relation (1.9) s'écrit (voir [3]) :

(1.11) $\mu [X, Y] = d \wedge \mu(X) \wedge d \mu(Y) + K_X \mu(Y)$

III.2 - <u>Algèbre de Lie dynamique sur une Variété Symplectique</u> : <u>Parenthèse de POISSON</u>.

Considérons $\mathcal{N} = C^{\infty}(M)$. Soient α et β des 1-formes sur la variété symplectique (M, F), X_α et X_β les champs des vecteurs correspondant respectivement.

2.1 - <u>Définition</u>: La parenthèse (ou crochet) de Poisson de α et β relativement à la structure symplectique de M est la 1-forme :

(2.2) $\{\alpha, \beta\} = i_{[X_\alpha, X_\beta]} F$

Muni de la parenthèse de Poisson l'espace $\Lambda^1(M)$ des 1-formes différentielles différentiables sur M est une algèbre de Lie.

Si α et β sont fermées on vérifie aisément que

(2.3) $\{\alpha, \beta\} = - d F (X_\alpha, X_\beta)$

Soient maintenant f et $g \in \mathcal{N}$, X_f et X_g les champs de vecteurs correspondants respectivement à df et dg.

2.4 - <u>Définition</u> : La parenthèse de Poisson (relativement à F) des fonctions f et g est la fonction différentitable :

(2.5) $\{ f, g \} = F (X_f, X_g) = X_f(g) = -X_g(f)$

Il résulte de considérations classiques (voir [4] , [5] que $\mathcal{N}$ muni de la parenthèse de Poisson est une algèbre de Lie, appelée algèbre de Lie dynamique sur la variété symplectique (M, F). L'algèbre de Lie dynamique $\mathcal{N}$ et l'algèbre associative $C^{\infty}(M)$ sont liées par la relation suivante :

(2.6) $\{ f, gh \} = h\{ f, g \} + g\{ f, h\}$

Soit $(q_1,\dots, q_n, p_1,\dots, p_n)$ un système de coordonnées locales sur un ouvert U de M tel que $F_{/U} = d p \wedge d q = \sum d p_i \wedge d q_i$. Alors dans U on a :

(2.7) $\{ f, g \} = \sum_{i=1}^{n} \frac{\partial f}{\partial p_i} \cdot \frac{\partial g}{\partial q_i} - \frac{\partial f}{\partial q_i} \cdot \frac{\partial g}{\partial p_i}$

(2.8) - Remarque : La parenthèse de Poisson peut également être définie de la manière suivante [4] :

Supposons qu'à $H \in \mathcal{N}$ correspond un groupe à un paramètre $\{\phi_t\}_{t \in \mathbb{R}}$ de transformations de M , notamment le flot de hamiltonien H. Soit $f \in \mathcal{N}$. La parenthèse de Poisson de f et H est définie par :

$$\{f, H\}(x) = \frac{d}{dt} f(\phi_t(x))\Big|_{t=0}, \forall x \in M ;$$

i.e. la dérivée de f suivant la direction du flot de hamiltonien H.
De cette remarque (2.8) découle la :

2.9 - Proposition : Une fonction $f \in \mathcal{N}$ est intégrale première du flot $\{\phi_t\}$ de Hamiltonien H si et seulement si $\{f, H\} = 0$.

On déduit aisément de la proposition (2.9) ci-dessus la généralisation suivante du théorème de E. NOETHER [4] .

2.10 - Théorème : Soient $H, f \in \mathcal{N}$ deux fonctions de Hamilton sur la variété symplectique (M, F). Si H est invariante par le groupe à un paramètre $\{\phi_t\}$ de transformations engendré par f alors f est intégrale première du système de hamiltonien H.

De l'identité de Jacobi relatif à $\mathcal{N}$ et compte tenu de la proposition (2.9) nous avons :

(2.11) - Théorème (Poisson) : Le crochet de Poisson (ou parenthèse de Poisson) $\{f, g\}$ de deux intégrales premières f et g d'un système de hamiltonien H est une intégrale première.

Ainsi d'après le théorème de Poisson les intégrales premières du flot de hamiltonien H engendre une sous-algèbre de Lie de l'algèbre de Lie $\mathcal{N}$. Enfin il est facile de vérifier que :

Si X_f et X_g sont des champs hamiltoniens de fonctions de Hamilton f et g respectivement alors $[X_f, X_g]$ a pour fonction de Hamilton $\{f, g\}$. Ainsi l'application de l'algèbre de Lie $\mathcal{N}$ sur l'algèbre de Lie L des champs hamiltoniens est un homomorphisme d'algèbres dont le noyau se compose de fonctions localement constantes (de fonctions constantes si M est connexe).

Si $\{\phi_f^{\,t}\}$ et $\{\psi_g^{\,t}\}$ sont deux flots de Hamiltoniens respectifs f et $g \in \mathcal{N}$

alors : $\phi_f^{\,t} \circ \psi_g^{\,t} = \psi_g^{\,t} \circ \phi_f^{\,t} \Longleftrightarrow \{f, g\} = 0$

(2.12) - Si M est connexe l'espace vectoriel des fonctions de $\mathcal{N}$ modulo les constantes additives est manifestement isomorphe à L^* et la structure d'algèbre de Lie induite sur $\mathcal{N}_{/\mathbb{R}}$ est définie par :

(2.13) $\{u,v\} = \{\bar{u}, \bar{v}\} = \Lambda(du \wedge dv) = \Lambda(d\bar{u} \wedge d\bar{v})$

où $\bar{u}$ et $\bar{v} \in \mathcal{N}_{/\mathbb{R}}$ sont les classes respectives de u et $v \in \mathcal{N}$. $\mathcal{N}_{/\mathbb{R}}$ muni de la parenthèse de Poisson est une algèbre de Lie isomorphe à l'algèbre de Lie L^* .

III.3 - Idéaux et Dérivations des Algèbres L, L_o, L^*, L_o^*, L_1 et L^c.

Nous supposerons dans ce paragraphe que (M, F) est une variété symplectique paracompacte connexe de dimension $2n$. Sur une telle variété, LICHNEROWICZ, AVEZ et DIAZ-MIRANDA ont étudié, grâce à des méthodes fournies par la structure presque kählérienne de M et à une généralisation d'un Lemme de CALABI [13], les ideaux, les dérivations de L, L_o, L^*, L_o^*, L_1, L^c ainsi que leur cohomologie de CHEVALLEY. Ils sont arrivés aux conclusions suivantes (cf [3]) :

(1) L^*, L_o et L_o^* sont des idéaux de L (découle de (1.9) de III.1)

(2) L, L^*, L_o, L_o^*, L_1 sont des idéaux de L^c (d'après (1.11) de III.2)

(3) $[L, L] = [L^*, L^*] = L^*$ (résultat obtenu dans le cas compact par Arnold)

$L/_{L^*}$ est abélienne et $\dim {}^L/_{L^*} = b_1(M)$, où $b_1(M)$ est le premier nombre de Betti de M pour la cohomologie à support compact.

(4) $\mathcal{N}_o$ et $\mathcal{N}_1$ sont des idéaux de $\mathcal{N}_1$ (découle de (2.13) de III.2)

(5) L_o^* et L_1 admettent L_1 comme idéal dérivé. De plus

$$L_1 = [L_1, L_o] = [L_o^*, L_o] = [L^*, L_o] = [L_1, L] = [L_o^*, L]$$

(6) L, L^*, L_o, L_o^*, L_1 et tous leurs idéaux sont semi-simples.

<u>Définition</u> : Une dérivation de l'algèbre de Lie L est un opérateur

$D : L \longrightarrow L$ telle que :

$$\forall X, Y \in L \ : \ D[X, Y] = [DX, Y] + [X, DY]$$

Notons en accord avec [3] D, D^*, $\mathcal{D}$, $\mathcal{D}_1$ et D^c les dérivations des algèbres de Lie L, L^*, $\mathcal{N}$, $\mathcal{N}_1$ et L^c respectivement.

Alors les principales conclusions sont :

(1) Toute dérivation de L^* (resp. de L) est une transformation infinitésimale conforme symplectique

$$Y \longrightarrow [X, Y] \quad , \text{où} \quad X \in L^c \ ;$$

(2) $\forall Y \in L^c \ : \ D^c Y = \mathcal{L}_X Y$, où $X \in L^c$,

c'est-à-dire que toute dérivation de L^c est intérieure.

(3) L'algèbre de Lie des dérivations locales de $\mathcal{N}$ est isomorphe à l'algèbre de Lie L^c par l'isomorphisme défini comme suit :

$$\mathcal{D}_X \equiv \mathcal{L}_X + K_X \ , \ \forall X \in L^c \ .$$

(4) Si M est compacte, toute dérivation de l'algèbre de Lie $\mathcal{N}$ est donnée par

$$\mathcal{D}\mu = \mathcal{L}_X \mu + V^{-1} \mathcal{D}(1) \int_M \mu \eta$$

où $X \in L$ et $\mathcal{D}(1) \in \mathbb{R}$, $V = \mathrm{vol}(M)$.

IV - SYSTEMES HAMILTONIENS DE DIMENSION INFINIE

IV.1 - Nous examinons dans ce paragraphe les systèmes hamiltoniens de dimension infinie, i.e. les systèmes hamiltoniens sur les espaces de Banach et les variétés banachiques.

Ces dernières années de nombreux résultats ont été obtenus sur les systèmes envisagés. La plupart des méthodes conduisant aux résultats sont fondées d'une part sur la théorie des semi-groupes introduites vers 1949 indépendant par HILLE et YOSIDA ([11]) (dans le cas des systèmes linéaires) et d'autre part sur des techniques introduites dans le cas non linéaire (semi-groupes sur les variétés banachiques) par BREZIS [7] , KATO [9] , SEGAL [8] (monotonicité, perturbations de LIPSCHITZ, systèmes hyperboliques symétriques), lesquelles sont liées à la recherche de solutions d'équations différentielles (non linéaires). Les résultats obtenus bien que importants sont encore loin d'être satisfaisants. Nous nous contenterons comme dans les deux premiers paragraphes (II et III) de dégager les principaux résultats. Pour les détails des démonstrations, se referer à [6] .

IV.2 - Systèmes Hamiltoniens Linéaires

Soit $\mathcal{B}$ un espace de Banach (réel ou complexe) et $F : \mathcal{B} \times \mathcal{B} \longrightarrow \mathbb{R}$ une forme symplectique sur $\mathcal{B}$. Pour tout $x \in \mathcal{B}$, l'espace tangent $T_x\mathcal{B}$ en x à $\mathcal{B}$ peut être identifié canoniquement à $\mathcal{B}$. Déterminons la 2-forme différentielle Ω sur $\mathcal{B}$ comme suit :

(1.1) $\Omega_x(e, \omega) = F(e, \omega)$, $\forall e, \omega \in \mathcal{B}$.

Il est clair que $d\Omega = 0$ car Ω_x est constante comme fonction de la variable x. Si $S \in \mathrm{End}(\mathcal{B})$ est telle que la dérivée $D_x S = S$ en x alors :

(1.2) $(S^* \Omega)_x (e, \omega) = \Omega_{S(x)} (Se, S\omega) = F(Se, S\omega)$.

Ainsi S est symplectique ($S^* \Omega = \Omega$) si et seulement si F est invariant par S ($S^* F = F$).

On sait que si $\phi_t = e^{tA}$ est un groupe (semi-groupe) de générateur A le domaine de définition $\mathcal{D}(A)$ de A est un sous-espace vectoriel dense de $\mathcal{B}$(cf [11]). Naturellement, à moins que A ne soit borné, ce champ de vecteurs est discontinu et non partout défini. Nous reviendrons ultérieurement sur les champs de vecteurs discontinus non linéaires. Concernant le cas linéaire qui est de loin le plus abordable nous avons le théorème suivant [6] :

(1.3) - <u>Théorème</u> : Soit $(\mathcal{B}, F)$ un espace de Banach faiblement symplectique et Ω la 2-forme différentielle correspondant à F . Soit $\{\phi_t\}$ le groupe à un paramètre (semi-groupe) engendré par A sur $\mathcal{B}$.

Alors les propositions suivantes sont équivalentes :

(1) $\mathcal{L}_A \Omega = 0$ (i.e. A est localement hamiltonien)

(2) $F(Ae, \omega) = -F(e, A\omega)$, $\forall e, \omega \in \mathcal{D}(A)$;

i.e. A est anti-symétrique par rapport à F ;

(3) A est globalement hamiltonien d'hamiltonien H_A défini par :

$$H_A(e) = \frac{1}{2} F(Ae, e), \forall e \in \mathcal{D}(A)$$

(4) $\{\phi_t\}$ est symplectique $(\phi_t^*(F) = F)$.

De plus sous les conditions ci-dessus on a : $H_A \circ \phi_t = H_A$ sur $\mathcal{D}(A)$ i.e. l'énergie est conservée.

Soit B un opérateur linéaire sur $\mathcal{B}$, de domaine $\mathcal{D}(B)$ dense. L'adjoint $B^\dagger$ de B relativement à F a pour domaine $\mathcal{D}(B^\dagger)$ défini par l'ensemble des $f \in \mathcal{B}$ auxquels correspond un $g \in \mathcal{B}$ tel que

$$F(Be, f) = F(e, g) \quad \forall e \in \mathcal{D}(B).$$

On pose $g = B^\dagger(f)$. On vérifie aisément que $B^\dagger$ est bien défini, linéaire et clos. Cela étant nous avons la version symplectique suivante du théorème de M. H. STONE [11] :

(1.4) - <u>Théorème</u> : Si la 2-forme symplectique F sur $\mathcal{B}$ est non dégénérée et si A engendre un groupe à un paramètre $\{\phi_t\}$ de transformations symplectiques sur $\mathcal{B}$ alors A est anti-adjoint $(A^\dagger = -A)$.

Cependant la réciproque du théorème (1.4) ci-dessus est fausse. Un contre-exemple nous est fourni par l'équation des ondes de KLEIN-GORDON (cf [6]). Un autre contre-exemple est fourni par l'équation de LAPLACE sur $\mathbb{R}^{n+1}$ (exemple de HADAMARD), concernant le problème de Cauchy "mal posé" $\frac{\partial^2 u}{\partial t^2} + \Delta u = 0$, où $B = -\Delta$.

En effet il n'existe réellement pas d'espace de Banach convenable sur lequel cette équation engendre un flot.

Soit maintenant A et B deux opérateurs linéaires anti-symétriques sur $\mathcal{B}$, d'hamiltoniens H_A et H_B respectivement. Le commutateur $[A, B] = AB - BA$ est anti-symétrique mais n'est cependant pas en général anti-adjoint, excepté le cas où A et B sont bornés ; en fait généralement le domaine de définition de $[A, B]$ n'est pas dense et $[A, B]$ n'est pas cloturable. Si A et B sont anti-symétriques relativement à F et $x \in \mathcal{D}([A, B])$ on a :

(1.5) $\{H_A, H_B\}(x) = H^{(x)}_{[A,B]}$.

En ce qui concerne la conservation de l'énergie pour les systèmes linéaires envisagés nous avons [6] :

(1.6) - <u>Théorème</u> : Soit $(\mathcal{B}, F)$ un espace de Banach faiblement symplectique, A et B deux opérateurs linéaires sur $\mathcal{B}$. Supposons que A et B engendrent respectivement sur $\mathcal{B}$ les groupes (semi-groupes) à un paramètre $\{\phi_t\}$ et $\{\psi_t\}$ d'hamiltonien H_A et H_B (respectivement). Si $\{\psi_t\}$ laisse $\mathcal{D}(A)$ invariant et $H_A \circ \psi_t = H_A$ alors $\{\phi_t\}$ laisse $\mathcal{D}(B)$ invariant et $H_B \circ \phi_t = H_B$; de plus $\phi_t \circ \psi_s = \psi_s \circ \phi_t$, $\forall t, s$.

Soit maintenant $\mathcal{H}$ un espace de Hilbert. Définissons la 2-forme symplectique F sur $\mathcal{H}$ par :

$$F(x, y) = \text{Im} < x, y > , \forall x, y \in \mathcal{H} .$$

On vérifie immédiatement que F est fortement non dégénérée. Nous avons alors les propriétés suivantes :

(1.7) (1) Une application linéaire $U : \mathcal{H} \longrightarrow \mathcal{H}$ est symplectique ($U^* F = F$) si et seulement si U est une isométrie.

(2) Si A est un opérateur linéaire complexe sur $\mathcal{H}$ de domaine $\mathcal{D}(A)$ dense dans $\mathcal{H}$ alors A est un champ de vecteurs hamiltoniens si et seulement si iA est symétrique.

De plus A engendre un flot si et seulement si iA est auto-adjoint. L'hamiltonien H_A de A est donné par la formule

$$H_A(x) = - \frac{1}{2} < i A x, x > , \forall x \in \mathcal{D}(A) .$$

Il convient de noter qu'une classe importante de systèmes hamiltoniens complexes nous est fournie par la mécanique quantique. Soit par exemple une particule non relativiste de masse m en mouvement dans un champ de force qui dérive d'une potentiel V(x). Alors $\mathcal{H} = \mathcal{L}^2 (\mathbb{R}^2 ; \mathbb{C})$ et

$$H = - \frac{1}{2m} \Delta + V(x) = - \frac{1}{2m} \left(\frac{\partial^2}{\partial x_1^2} + \frac{\partial^2}{\partial x_2^2} + \frac{\partial^2}{\partial x_3^3} \right) + V(x) .$$

Il est très important d'établir des théorèmes précis, moyennant des conditions sur V , qui garantissent qu'à l'expression ci-dessus correspond un opérateur auto-adjoint unique tel qu'un groupe à un paramètre bien défini $\phi_t = e^{itH}$ existe. KATO [9] a montré que les hamiltoniens usuels des physiques atomique et moléculaire non relativistes sont essentiellement auto-adjoints.

IV.3 - Systèmes Hamiltoniens Sur Une Variété Banachique

Soit M une variété banachique (i.e. modelée sur un espace de Banach $\mathcal{B}$) et D un sous-ensemble de M .

(2.1) - Définition : Un flot sur D est une collection $\{\phi_t\}$ d'applications $\phi_t : D \longrightarrow D$ définie $\forall t \in \mathbb{R}$ telles que :

(1) $\phi_o = Id_D$

(2) $\phi_{t+s} = \phi_t \circ \phi_s \ , \forall t, s \in \mathbb{R}$

Notons que $\phi_{-t} \circ \phi_t = Id_D$ et $\phi_t \circ \phi_{-t} = Id_D$. Ainsi ϕ_t est bijective et $\phi_t^{-1} = \phi_{-t}$.

Un semi-flot sur D est une collection d'applications $\phi_t : D \longrightarrow D$, définies pour $t \geqslant 0$ satisfaisant à (1) et (2) ci-dessus.

(2.2) - Définition : Un champ de vecteurs X avec domaine de définition, D est une application $X : D \longrightarrow T(M)$ telle que $\forall x \in D$, $X(x) \in T_x M$.

Une courbe intégrale de X est une application

$$\gamma :]a, b[\subset \mathbb{R} \longrightarrow D ,$$

différentiable en tant qu'application à valeurs dans M qui vérifie :

$$\frac{d\gamma}{dt} = X_{\gamma(t)}$$

Un flot sur D tel que $\forall x \in D$ l'application $t \longrightarrow \phi_t(x)$ est une courbe intégrale de X est dit flot engendré par X sur D (les semi-flots engendrés par X sont définis de manière analogue).

Si $\phi : \mathbb{R} \times M \longrightarrow M$ est un flot tel que ϕ soit de classe C^o (en tant que fonction de 2 variables) alors $\{\phi_t\}_{t \in \mathbb{R}}$ est dit flot de classe C^o sur M . Si le flot $\{\phi_t\}$ de X est la restriction à D d'un flot de classe C^o sur M alors $\{\phi_t\}$ est appelé flot de classe C^o de X .

Soit $\{\phi_t\}_{t \in \mathbb{R}}$ un flot de classe C^o sur M . Si pour tout t fixé, l'application $\phi_t : M \longrightarrow M$ est de classe C^k alors $\{\phi_t\}$ est dit flot de classe C^k .

(2.3) - Remarque : Un flot est dit de classe C^k si ϕ_t est de classe C^k en la variable x mais non nécessairement en la variable t . Nous verrons (Théorème (2.5)) qu'un flot n'est de classe C^k en la variable t que dans le seul cas où il est engendré par champ de vecteurs partout défini.

Soit $\{\phi_t\}$ un flot sur la variété banachique CHERNOFF et MARSDEN [6] ont montré que :

(2.4) - Si les applications $t \longrightarrow \phi_t(x)$ et $x \longrightarrow \phi_t(x)$ sont continues respectivement pour x et t fixés alors $\phi : \mathbb{R} \times M \longrightarrow M$ est de classe C^o , i.e. le flot $\{\phi_t\}$ est de classe C^o.

D'après la remarque (2.3) ci-dessus un flot $\{\phi_t\}$ sur la variété banachique M ne sera de classe C^k en la variable t que s'il est engendré par un champ de vecteurs partout défini. KATO [9] a donné des exemples de générateurs X qui ne possèdent pas de flots différentiables. Concernant donc ce problème d'existence de champ de vecteurs partout définis nous avons la généralisation suivante d'un résultat classique dans le cas linéaire (cf [6]) :

2.5 - Théorème : Soit $\phi : \mathbb{R} \times M \longrightarrow M$ un flot C^o sur la variété de Banach M . Supposons que pour tout t, ϕ_t est de classe C^k, $k \geqslant 1$ et que $\forall x \in M$,

$||D\, \phi_t(x) - I|| \to 0$ quand $t \to 0$, où $||.||$ désigne la norme de l'opérateur. Alors $\phi : \mathbb{R} \times M \longrightarrow M$ est de classe C^k. De plus le générateur X de $\{\phi_t\}$ est partout défini et de classe C^{k-1} sur M.

2.6 - Définition : Soit $\{\phi_t\}$ un flot sur une variété banachique M . Le générateur de $\{\phi_t\}_{t\in\mathbb{R}}$ est le champ de vecteurs X défini par

$$X(x) = \frac{d}{dt}\,\phi_t(x)\Big|_{t=0} ;$$

le domaine de définition de X consiste donc en les points $x \in M$ tels que la limite indiquée ci-dessus existe.

Nous avons [6] le :

2.7 - Théorème : Soit $\{\phi_t\}$ un flot de classe C^1 , de générateur X . Alors ϕ_t applique D(X) dans lui-même et $t \longrightarrow \phi_t(x)$ est de classe C^1 pour $x \in D(X)$.

De plus on a :

$$X(\phi_t(x)) = T\,\phi_t(x)\ (X_x)$$

et X est un champ de vecteurs fermé, i.e. son graphe est fermé.

Il est tout à fait possible qu'un flot de classe C^o, $\{\phi_t\}$, ne possède pas de générateur (même dans le cas de la dimension finie).

Cependant KOMURA [10] a montré que si ϕ_t est C^o et lipschitzien avec $||\phi_t||_{Lip} \leqslant e^{\beta t}$ alors le générateur X de $\{\phi_t\}$ est défini et D(X) est dense dans M . Ainsi en accord avec KOMURA nous nous interrogeons de savoir dans quelles conditions un flot $\{\phi_t\}$ admet un générateur X bien défini et de domaine D(X) dense dans M . La réponse est fournie par le (cf [6]).

2.8 - Théorème : Soit $\{\phi_t\}$ un flot de classe C^2 sur M .
Soit X le générateur de $\{\phi_t\}$ et Y le générateur de $\{T\,\phi_t\}$. Alors pour $x \in D(X)$, l'ensemble $D_x = D(Y) \cap T_x M$ est un sous-espace linéaire dans $T_x M$.

Soit maintenant $X : D \longrightarrow TM$ un champ de vecteurs de domaine D sur la variété banachique M . Si X possède un flot C^o $\{\phi_t\}$ avec $\phi_t : D \longrightarrow D$ et si $f : M \longrightarrow \mathbb{R}$ est une fonction différentiable sur M alors, pour tout $x \in D$ on a :

$$(2.9)\quad \frac{d}{dt}\, f(\phi_t(x)) = \phi_t^*\,(X(f)(x)) ,$$

où $X(f) : M \longrightarrow \mathbb{R}$ est définie par $(X(f))(x) = d\,f(x)(X(x))$ et $\phi_t^*(f) = f \circ \phi_t$.

Pour établir des formules analogues à (2.9) ci-dessus pour les k-formes aussi bien que pour les fonctions scalaires sur une variété de Banach nous sommes conduits à introduire les définitions suivantes :

(2.10) - Définition : Un sous-ensemble D d'une variété banachique M est un domaine de variété si :

(1) D est dense dans M

(2) D possède une structure de variété de Banach (qui lui est propre) telle que l'inclusion $i : D \longrightarrow M$ soit différentiable.

(3) $\forall x \in D$, l'application linéaire $T_x i : T_x D \longrightarrow T_x M$ est une inclusion dense.

(exemples : les sous-espaces linéaires denses D d'un espace de Banach $\mathcal{B}$ possèdent une norme plus forte que celle de $\mathcal{B}$).

Si $X : D \longrightarrow M$ est un champ de vecteurs avec domaine D et α une k-forme sur M, on définit le produit intérieur $i_X \alpha$ par :

$$(i_X \alpha)_x \; (v_1,\dots, v_{k-1}) = \alpha_x \; (X(x), v_1,\dots, v_{k-1})$$

où $x \in D$ et $v_1,\dots, v_{k-1} \in T_x(D) \subset T_x(M)$.

Comme D possède une structure de variété on peut dès lors parler de champ de vecteurs de classe C^r , $X : D \longrightarrow TM$. Si $r \geq 1$ et X de classe C^r, de domaine D , α une k-forme de classe C^r sur M alors la dérivée de Lie est donnée par :

$$\mathcal{L}_X \alpha = d\, i_X \alpha + i_X \, d\alpha$$

Le flot $\{\phi_t\}$ de X est dit de classe C^1 sur le triplet $D_1 \subset D \subset M$ si :

(2.11) (1) $D_1 \subset D$; $D \subset M$ sont des domaines de variété

(2) $X : D \longrightarrow T(M)$ est un champ de vecteurs de flot C^o, $\phi_t : D \longrightarrow D$.

(3) $\{\phi_t\}$ laisse D_1 invariant, est C^o sur D_1 ;

(4) $\phi_t : D_1 \longrightarrow D$ est de classe C^1 avec $D\phi_t . v$ continue en tant que fonction des 2 variables (t, v).

Nous avons dans [6] le

(2.12) - <u>Théorème</u> : Soit $\{\phi_t\}$ un flot de classe C^1 du champ de vecteurs X sur le triplet $D_1 \subset D \subset M$. Si $X : D \longrightarrow TM$ est de classe C^1 et α une k-forme de classe C^1 sur M on a la relation

$$\frac{d}{dt} \phi_t^* (\alpha) = \phi_t^* (\mathcal{L}_X \alpha),$$

où les deux membres de l'égalité sont considérés comme des formes sur D_1.

Sous les hypothèses du Théorème (2.10) ci-dessus, $\mathcal{L}_X \alpha = 0$ entraîne $\phi_t^* (\alpha) = \alpha, \forall t$. En particulier si $d\alpha = 0$ alors $\phi_t^* (\alpha) = \alpha$ si $i_X \alpha$ est fermé ($d\, i_X \alpha = 0$).

(2.13) - <u>Définition</u> : Soit (M, F) une variété (banachique) faiblement symplectique. Un champ de vecteurs X de domaine de variété $D \subset M$ est dit hamiltonien s'il existe $H : D \longrightarrow \mathbb{R}$ telle que $i_X F = dH$.
X est dit localement hamiltonien si $d\, i_X F = 0$.
Nous écrivons X_H pour X .

La 2-forme F étant faiblement symplectique il n'existe pas nécessairement de champ de vecteurs X_H correspondant à H donnée sur D . De plus même si H est définie et différentiable sur la variété M tout entière, X_H n'est en général défini que sur un sous-ensemble de M . Il est cependant défini de manière unique par la condition ci-dessus.

Pour $x \in D$, $v \in T_x D \subset T_x (M)$, d'après les conditions ci-dessus on a :

$$F_x \, (X_H(x), v) = dH(x)(v).$$

<u>Définition</u> : Soit $X_f : D_1 \longrightarrow TM$ et $X_g : D_2 \longrightarrow TM$ deux champs de vecteurs hamiltoniens d'hamiltoniens respectifs f et g.
Le crochet (ou parenthèse) de Poisson est défini par :

$$\{f, g\} \ : \ D_1 \cap D_2 \longrightarrow \mathbb{R}$$

$$\{f, g\}(x) = F(x)\ (X_f(x), X_g(x), X_g(x))$$

$$\forall x \in D_1 \cap D_2 .$$

Soit X_H un champ de vecteurs hamiltoniens d'hamiltoniens h et de domaine D. Supposons que X_H engendre un flot C^o, $\phi_t : D \longrightarrow D$ sur D. Soit $f : M \longrightarrow \mathbb{R}$ une fonction de classe C^1.

Si $\{H, f\} = df.(X_H) = 0$ alors $f \circ \phi_t = f$, c'est-à-dire f est constante sur les trajectoires du flot de hamiltonien H. De manière plus générale nous avons le théorème suivant concernant la conservation de l'énergie ([6]) :

<u>Théorème</u> : Soit (M, F) une variété faiblement symplectique, $X_H : D \longrightarrow TM$ un champ de vecteurs hamiltonien de domaine variété $D \subset M$. Supposons que X_H engendre un flot C^o, $\phi_t : D \longrightarrow D$. Soit $f : D \longrightarrow \mathbb{R}$ une fonction de classe C^1.

Si à f est associé un champ de vecteurs hamiltoniens continu

$X_f : D \longrightarrow TM$ alors on a :

$$\frac{d}{dt}(f \circ \phi_t) = \{f, H\} \circ \phi_t \quad \text{sur } D .$$

En particulier si $\{f, H\} = 0$ alors $f \circ \phi_t = f$ sur D ./.

REFERENCES

[1] A. LICHNEROWICZ : Choix d'Oeuvres Mathématiques HERMANN (PARIS, 1982)

[2] A. LICHNEROWICZ : C.R. Acad. Sci. PARIS 233 (1951) p. 723-726

[3] Journ. Diff. Geometry Vol. 9 N° 1 (1974) 1-40

[4] V. ARNOLD : Bases Mathématiques de la Mécanique Classique. Ed. MIR, MOSCOU (1976).

[5] C. GODBILLON : Géométrie Différentielle et Mécanique Analytique. Ed. HERMANN (PARIS, 1976).

[6] PAUL R. CHERNOFF et J. E. MARSDEN : Properties of Infinite Dimensional Systems Lecture Notes 425 Springer Verlag (1974).

[7] H. BREZIS : Opérateurs maximaux monotones et contraction dans les espaces de Hilbert. NORTH-HOLLAND (1973).

[8] I. SEGAL : Ann. Math. 78 (1963) 339-364.

[9] T. KATO : Journ. Math. Soc. JAPAN 25 (1973) 648-660.

[10] Y. KOMURA : Journ. Math. Soc. JAPAN 19 (1967) 493-507.

[11] K. YOSIDA : Functional Analysis SPRINGER - VERLAG. (1971).

[12] S. LANG : Differentiable Manifolds. ADDISON-WESLEY (1964).

[13] E. CALABI : Problems in Analysis - Symp. in honour of S. BOCHNER Princeton Univ. Press (1970). p. 1-16.

HARMONIC SECTIONS, YANG-MILLS FIELDS, AND EINSTEIN'S EQUATION

C.M.WOOD

Department of Pure Mathematics, University of Liverpool,

Liverpool, Great Britain.

The objective of these talks is threefold:

(1) To show how the theory of connections in fibre bundles may be extended to foliations.

(2) To introduce the concept of a harmonic map. More precisely, and in tune with the title of this meeting, we shall be dealing with harmonic sections.

(3) To show how certain physically significant variational principles (Yang-Mills theory, and gravitation) may be used to construct and characterize harmonic sections.

For a more expansive treatment of these ideas, we refer to [7],[8], and [9].

Our starting point is a pair of observations from gauge theory, leading to a natural question.

§1 Gauge Theory

Let G be a Lie group with Lie algebra $\mathfrak{g}$, and let $\rho: M \to M'$ be a principal G-bundle with connection 1-form $\omega: TM \to \mathfrak{g}$. Suppose g' to be a Riemannian metric on M', and that G admits a bi-invariant metric <,> .

Then, the Kaluza-Klein metric g on M is defined as $g = \rho^* g' + \omega^* \langle , \rangle$. We have the exterior covariant derivative $D = D^{\omega}$, and its adjoint D^*. The curvature 2-form is defined $\Omega = D\omega$, and the two equations $D\Omega = 0$ and $D^*\Omega = 0$ are Bianchi's identity and Yang-Mills equation, respectively. We define distributions $\mathcal{F} = \ker d\rho$, and $\mathcal{F}^{\perp} = \ker \omega$, and denote by $\pi, \pi^{\perp}$ the corresponding orthogonal projections. The usual terminology and notation for $\ker d\rho$ and $\ker \omega$, of vertical $\mathcal{V}$ and horizontal $\mathcal{H}$, will be reserved for later use - see §2 below.

Observation 1 $\mathcal{F}$ is a totally geodesic (hereafter abbreviated to t.g.) Riemannian foliation of (M,g) [6].

Observation 2 We define the integrability 2-form (of $\mathcal{F}^{\perp}$) to be

$$I_{\mathcal{F}^{\perp}} = I : \mathcal{F}^{\perp} \times \mathcal{F}^{\perp} \to \mathcal{F} \ ; \quad I(X,Y) = \tfrac{1}{2}\pi[X,Y] \ .$$

Then, $\omega \circ I = -\frac{1}{2}\Omega$; or, equivalently, $I = -\frac{1}{2}\Omega^*$, where $\Omega^*(X,Y) = \frac{d}{dt}\Big|_0 x . \exp t\Omega(X,Y)$, for any $X,Y \in T_xM$.

Question Suppose that $\mathcal{F}$ is any t.g. Riemannian foliation of (M,g) - not necessarily a fibration - with integrability 2-form I. Does there exist a potential for I ? (Answer See Theorem 2·1 below.)

§2 The Gauss Section

For any integer k, $0 < k < m = \dim M$, there is the Grassmann bundle $\xi : G_kM \to M$ of k-planes in TM . Then, any k-dimensional distribution $\mathcal{F} \subset TM$ defines, and is defined by, a section $\gamma = \gamma_{\mathcal{F}}$ of ξ - the Gauss section. The Levi-Civita connection of a Riemannian metric g on M gives a splitting $TG_kM = \mathcal{V} \oplus \mathcal{H}$, where $\mathcal{V} = \ker d\xi$. In order to relate the geometrical properties of γ to those of $\mathcal{F}$, we define the following Lie algebra of nilpotent matrices, and associated vector bundle of nilpotent endomorphisms of TM :

$$\mathfrak{a} = \left\{ \begin{pmatrix} 0_k & Y \\ 0 & 0_{m-k} \end{pmatrix} \right\}$$

$$\mathfrak{a}_{TM} = \{ A \in \text{End TM} \mid \text{im}(A) \subset \mathcal{F} \text{ and } \ker(A) \supset \mathcal{F} \} .$$

There is then an exceptionally nice isomorphism $\kappa : \gamma^{-1} \mathcal{V} \to \mathfrak{a}_{TM}$; see [7]. Speaking heuristically, κ may be viewed as the following composite:

$$\mathfrak{a}_{TM} \cong \mathcal{F}^* \otimes \mathcal{F}^{\perp} = \gamma^{-1}K^* \otimes \gamma^{-1}K^{\perp} = \gamma^{-1}(K^* \otimes K^{\perp}) \cong \gamma^{-1}\mathcal{V}$$

where $K, K^{\perp} \to G_k M$ are the universal vector bundles.

Now, let $(d\gamma)^{\mathcal{V}}$ denote the $\mathcal{V}$-component of the differential.

<u>Theorem 2·1</u> [9]

Suppose that $\mathcal{F}$ is a t.g. Riemannian foliation of (M,g) . Then

$$\kappa \circ (d\gamma)^{\mathcal{V}} = -I .$$

In analogy with the formation of the Kaluza-Klein metric, there is a Riemannian metric h on $G_k M$ - a combination of the $\mathcal{H}$-lift of g with the canonical Grassmannian metric on $\mathcal{V}$. Denote by ${}^{\mathcal{V}}\nabla$ the $\mathcal{V}$-component of the Levi-Civita connection of h . The <u>vertical tension field</u> of γ is then defined

$$\tau^{\mathcal{V}}(\gamma) = \text{Trace}_h \, {}^{\mathcal{V}}\nabla (d\gamma)^{\mathcal{V}} .$$

If $\tau^{\mathcal{V}}(\gamma) = 0$, we say that γ is a <u>harmonic section</u>.

<u>Theorem 2·2</u> [9]

Let $\mathcal{F}$ be a t.g. Riemannian foliation. Then

$$\kappa \circ \tau^{\mathcal{V}}(\gamma) = d^* I .$$

<u>Note</u> The exterior coderivative d^* is that for $\mathcal{F}$-valued differential forms on $\mathcal{F}^{\perp}$, and is built from the appropriate projections of the Levi-Civita

connection of (M,g). The same goes for d, in Lemma 3·1 below.

In summary, for a t.g. Riemannian foliation $\mathcal{F}$ we have the following bicorrespondences, those on the right being for the special case when $\mathcal{F}$ is obtained from a principal bundle with connection:

$(d\gamma)^{V}$	$\longleftrightarrow$	$-I$	$\longleftrightarrow$	Ω
$\tau^{V}(\gamma)$	$\longleftrightarrow$	$d^{*}I$	$\longleftrightarrow$	$-D^{*}\Omega$
Harmonic γ	$\longleftrightarrow$	Co-closed I	$\leftarrow\cdots\rightarrow$	Yang-Mills ω

Properly speaking, $\leftarrow\cdots\rightarrow$ is a bicorrespondence only after identifying connections under the action of the automorphism group of the bundle. We shall return to examine the correspondence between harmonic γ and Yang-Mills ω from a variational viewpoint. But first some special cases.

§3 Special Cases

(A) The t.g. Riemannian foliations are notable in admitting a generalization of Bianchi's identity.

Lemma 3·1 [9]

Let $\mathcal{F}$ be a t.g. Riemannian foliation. Then $dI = 0$.

On the algebra of $\mathcal{F}$-valued differential forms on $\mathcal{F}^{\perp}$, there is a Hodge * operator, taking p-forms to (m-k-p)-forms; $\alpha \mapsto *\alpha$. The coderivative d^{*} is then simply $*d*$. In case $m-k = 4$, say that γ (or $\mathcal{F}$) is ±self-dual if $*I = \pm I$.

Theorem 3·2

Let $\mathcal{F}$ be a t.g. Riemannian foliation. If γ is ±self-dual, then γ is harmonic.

Proof It follows from Theorem 2·2 and Lemma 3·1 that

$$\kappa \circ \tau^{V}(\gamma) = d^{*}I = *d*I = \pm * dI = 0 .$$

(B) Denote by h^{V} the degenerate quadratic differential on $G_k M$

$$h^{V}(X,Y) = h(X^{V},Y^{V}) .$$

Then γ is weakly conformal if $\gamma^{*}h^{V} = f.g$, for some $f: M \to \mathbb{R}^{+} \cup \{0\}$.

Theorem 3·3 [9]

Let $\mathcal{F}$ be a t.g. Riemannian foliation. Then $\gamma^{*}h^{V}(\mathcal{F},\mathcal{F}) = 0 = \gamma^{*}h^{V}(\mathcal{F},\mathcal{F}^{\perp})$. Moreover, in case m-k = 2

$$\gamma^{*}h^{V}(\mathcal{F}^{\perp},\mathcal{F}^{\perp}) = \{ \overline{K}(\mathcal{F}^{\perp}) - K(\mathcal{F}^{\perp}) \} g$$

where $\overline{K}$ denotes the intrinsic sectional curvature, and K the extrinsic sectional curvature. In particular, γ is weakly conformal.

§4 The Yang-Mills and Vertical Energy Functionals

In this section, let M be the total space of a principal bundle with connection, assumed compact for simplicity. The Yang-Mills energy of the connection ω is defined

$$\mathcal{YM}(\omega) = \tfrac{1}{2}\int_M ||\Omega||^2 \, \mathrm{vol}(g)$$

and $D^{*}\Omega = 0$ precisely when ω is a critical point of $\mathcal{YM}$. On the other hand, the vertical energy of γ is defined

$$E^{V}(\gamma) = E^{V}(\gamma,g) = \tfrac{1}{2}\int_M ||(d\gamma)^{V}||^2 \, \mathrm{vol}(g)$$

and $\tau^{V}(\gamma) = 0$ precisely when γ is a critical point of E^{V} with respect to variations through sections.

There is the evident duality $\gamma \leftrightarrow \gamma^{\perp}$, where $\gamma^{\perp}$ is the Gauss section of $\mathcal{F}^{\perp}$. Then $E^{V}(\gamma) = E^{V}(\gamma^{\perp})$, and the two variational problems are entirely equivalent. The Yang-Mills energy is a functional naturally suited to the "analytic" view of connections as differential forms, whereas the vertical energy suits the more geometric viewpoint of a connection as a plane field complementary to the fibres.

It is clear that any 1-parameter variation ω_t of ω generates 1-parameter variations g_t of g and $\gamma_t^{\perp}$ of $\gamma^{\perp}$. We then have (Theorem 2·1):

$$\mathcal{YM}(\omega_t) = E^{V}(\gamma, g_t) = E^{V}(\gamma_t^{\perp}, g_t) .$$

Since the harmonic $\gamma, \gamma^{\perp}$ are those critical points of E^{V} with respect to a fixed metric, it is not immediately obvious that the two variational principles $\mathcal{YM}$ and E^{V} are equivalent. Our objective here is to demonstrate how to arrive at Theorem 2·2 from the variational direction.

The space of connection forms in $\rho: M \to M'$ is an affine space modelled on the following subspace of $\mathfrak{g}$-valued 1-forms:

$$\mathfrak{A}^1(M;\mathfrak{g}) = \{ A: TM \to \mathfrak{g} \mid A(\mathcal{F}) = 0 \text{ , } A \text{ is G-equivariant} \} .$$

A natural variation of ω is therefore $\omega_t = \omega + tA$. Now $\mathfrak{A}^1(M;\mathfrak{g})$ is a Lie algebra. To exhibit $\mathfrak{A}^1(M;\mathfrak{g})$ as the Lie algebra of a Lie group, we recall the bundle $\mathfrak{a}_{TM} \to M$ and write $\mathfrak{A}_{TM} = C(\mathfrak{a}_{TM})$, the space of sections. Since the fibres of $\mathfrak{a}_{TM}$ are Lie algebras, $\mathfrak{A}_{TM}$ is a Lie algebra, and contains $\mathfrak{A}^1(M;\mathfrak{g})$ as the following subalgebra:

$$\mathfrak{A}^1(M;\mathfrak{g}) \hookrightarrow \mathfrak{A}_{TM} \text{ ; } A \mapsto A^*$$

where $A^*(X) = \frac{d}{dt}\Big|_0 x.\exp tA(X)$, for any $X \in T_xM$. Now define

$$A_{TM} = \{ \phi \in \text{Aut}\, TM \mid \phi = 1 + A \text{ , } A \in \mathfrak{a}_{TM} \}$$

$$\mathcal{A}_{TM} = C(A_{TM}) .$$

Since the fibres of $A_{TM} \to M$ are Lie groups, $\mathcal{A}_{TM}$ is a Lie group. Its Lie algebra is $\mathfrak{A}_{TM}$, and $\mathfrak{A}^1(M;\mathfrak{g})$ is the Lie algebra of the subgroup of G-equivariant elements. Now, $\mathcal{A}_{TM}$ acts as follows:

$$\phi.\omega = (\phi^{-1})^*\omega = \omega - A \text{ , provided } A \in \mathfrak{A}^1(M;\mathfrak{g})$$

$$\phi.g = (\phi^{-1})^* g$$

$$\phi.\gamma^\perp = \gamma_{\phi(\mathcal{F}^\perp)} \quad .$$

Notice that $\phi.\gamma = \gamma$, since $\phi|\mathcal{F} = id_{\mathcal{F}}$. To get an action on γ we introduce the group $\mathcal{A}^\dagger_{TM}$, the adjoint of $\mathcal{A}_{TM}$ with respect to the metric g. Let $\psi \in \mathcal{A}^\dagger_{TM}$ denote the adjoint of $\phi \in \mathcal{A}_{TM}$. The actions of $\mathcal{A}^\dagger_{TM}$ on γ and $\mathcal{A}_{TM}$ on $\gamma^\perp$ are then dual with respect to E^V, in the following sense:

$$E^V(\psi.\gamma,\psi.g) = E^V(\phi.\gamma^\perp,\phi.g) \quad .$$

Now, let $\phi_t = 1 + tA$ be a 1-parameter subgroup of $\mathcal{A}_{TM}$, with $A \in \mathfrak{A}^1(M;\mathfrak{g})$. We have

$$\mathcal{YM}(\phi_t.\omega) \ (= \mathcal{YM}(\omega_{-t})\ ) = E^V(\gamma,\phi_t.g) = E^V(\phi_t.\gamma^\perp,\phi_t.g)$$

$$= E^V(\psi_t.\gamma,\ \psi_t.g) \quad .$$

So $$\left.\frac{d}{dt}\right|_o \mathcal{YM}(\phi_t.\omega) = \left.\frac{d}{dt}\right|_o E^V(\phi_t.\gamma,g) + \left.\frac{d}{dt}\right|_o E^V(\gamma,\phi_t.g) \quad .$$

In analogy with [1,5] we define the <u>vertical stress energy tensor</u> to be

$$S^V(\gamma,g) = \tfrac{1}{2}\,\|(d\gamma)^V\|^2 g - \gamma^* h^V \quad .$$

Then, for any variation g_t of g, we have

$$\left.\frac{d}{dt}\right|_o E^V(\gamma,g_t) = \tfrac{1}{2}\int_M g(\,S^V(\gamma,g), \left.\frac{d\,g_t}{dt}\right|_o)\ vol(g)$$

Now, $\left.\dfrac{d\,\phi_t.g}{dt}\right|_o = -(A + A^\dagger)$, a quadratic differential vanishing on $\mathcal{F}\times\mathcal{F}$ and $\mathcal{F}^\perp\times\mathcal{F}^\perp$. By Theorem 3·3, $\gamma^* h^V(\mathcal{F},\mathcal{F}^\perp) = 0$. We conclude that

$$\left.\frac{d}{dt}\right|_o \mathcal{YM}(\phi_t.\omega) = \left.\frac{d}{dt}\right|_o E^V(\psi_t.\gamma,g) \quad .$$

In order to deduce that the first variations of $\mathcal{YM}$ and E^{ν} coincide, when we have only shown that to be true for G-equivariant variations of γ, we need to appeal to the Principle of Symmetric Criticality [4]. See also Note 6·3.

Our aim now is to use the foregoing variational technique in conjunction with Einstein's equation, to derive a more geometric realization of $\tau^{\nu}(\gamma)$ when $\mathcal{F}$ is a t.g. Riemannian foliation.

§5 The Kaluza-Klein Curvature Identity

For a principal fibre bundle $\rho: M \to M'$ with connection, we recall the following decomposition of the scalar curvature of the Kaluza-Klein metric [2]:

$$S = S' \circ \rho + S_{\rho} - \tfrac{1}{4}\|\Omega\|^2$$

where S_{ρ} is the scalar curvature of the fibres (= constant). Now, suppose that $\mathcal{F}$ is a foliation of M. Let $\rho: TM \to TM/\mathcal{F}$ denote the quotient morphism, and define the adapted connection $\tilde{\nabla}$ in $TM/\mathcal{F} \to M$ as follows:

$$\tilde{\nabla}_X(\nu Y) = \begin{cases} \nu(\nabla_X Y), & \text{if } X \in \mathcal{F}^{\perp} \\ \nu[\pi X, Y], & \text{if } X \in \mathcal{F}. \end{cases}$$

Let $\tilde{R}$ denote the curvature of $\tilde{\nabla}$ - the adapted curvature. Let $\{E_i : 1 \leq i \leq m\}$ be an $\mathcal{F}$-adapted (or Darboux) orthonormal frame; that is, $\{E_1, \cdots, E_k\} \subset \mathcal{F}$ and $\{E_{k+1}, \cdots, E_m\} \subset \mathcal{F}^{\perp}$. The adapted scalar curvature is then defined

$$S_{\nu} = \sum_{i,j>k} \tilde{g}(\tilde{R}(E_i, E_j,)\, \nu E_j, \nu E_i) = S_{\nu}(g)$$

where $\tilde{g}$ is the metric induced by the isomorphism $TM/\mathcal{F} \to \mathcal{F}^{\perp}$. The $\mathcal{F}$-partial scalar curvature is

$$S_{\mathcal{F}} = \sum_{i,j \leq k} g(R(E_i, E_j)\, E_j, E_i) = S_{\mathcal{F}}(g)$$

where R is the Riemann curvature tensor of (M,g). If $\mathcal{F}$ is totally geodesic, then Gauss' equation implies that $S_{\mathcal{F}}$ is the scalar curvature of the leaves, in their induced metric.

Theorem 5·1 [8]

Let $\mathcal{F}$ be a t.g. Riemannian foliation of (M,g). Then

$$S = S_{\nu} + S_{\mathcal{F}} - \|(d\gamma)^{\mathcal{V}}\|^2 .$$

§6 Einstein's Equation

Integrating Theorem 5·1 yields

$$I(g) = \int_M (S_{\nu}(g) + S_{\mathcal{F}}(g))\,\mathrm{vol}(g) - 2E^{\mathcal{V}}(\gamma,g)$$

where $I(g)$ is the integrated scalar curvature, or gravitational energy of g. As in §4, we have the groups A_{TM} and $A^{\dagger}_{TM}$, acting on g, $\gamma^{\perp}$, and γ. We note the following Stability Theorem.

Theorem 6·1 [8]

Suppose that $\mathcal{F}$ is a t.g. Riemannian foliation of (M,g), and let $\phi \in A_{TM}$. Then:

(i) $\mathcal{F}$ is a t.g. Riemannian foliation of $(M,\phi.g)$.

(ii) $S_{\nu}(\phi.g) = S_{\nu}(g)$.

(iii) $S_{\mathcal{F}}(\phi.g) = S_{\mathcal{F}}(g)$.

(iv) $\mathrm{vol}(\phi.g) = \mathrm{vol}(g)$.

We conclude that

$$\frac{d}{dt}\Big|_o I(\phi_t.g) = -2\frac{d}{dt}\Big|_o E^{\mathcal{V}}(\gamma,\phi_t.g) = -2\frac{d}{dt}\Big|_o E^{\mathcal{V}}(\psi_t.\gamma,g) .$$

The first variation of $I(g)$ is given by Einstein's equation [3]:

$$\left.\frac{d}{dt}\right|_o I(g_t) = \int_M g\left(\tfrac{1}{2} S g - \mathrm{Ric}, \left.\frac{d g_t}{dt}\right|_o\right) \mathrm{vol}(g)$$

for any variation g_t of g, where Ric denotes the Ricci curvature of (M,g).

Thus

$$\left.\frac{d}{dt}\right|_o I(\phi_t \cdot g) = 2\int_M g(\mathrm{Ric}(\mathcal{F},\mathcal{F}^\perp), A + A^\dagger)\, \mathrm{vol}(g) \quad .$$

Theorem 6·2 [7,8]

Let $\mathcal{F}$ be a t.g. Riemannian foliation of (M,g). Then

$$\tau^V(\gamma) = 0 \iff \mathrm{Ric}(\mathcal{F},\mathcal{F}^\perp) = 0 \ .$$

Note 6·3 To give a watertight proof of 6.2, we need to show that $\gamma_t = \psi_t \cdot \gamma$ is a sufficiently large class of variations to exhaust the "tangent space of γ".

Corollary 6·4

If in addition (M,g) is an Einstein space, then γ is harmonic.

REFERENCES

[1] P.BAIRD & J.EELLS, A conservation law for harmonic maps, Geometry Symposium Utrecht 1980, 1 - 25, Lecture Notes in Mathematics 894 (Springer, Berlin, 1981).

[2] D.BLEECKER, Gauge Theory & Variational Principles, Adison-Wesley, 1981.

[3] S.HAWKING & G.ELLIS, The Large Scale Structure of Space-Time, Cambridge University Press, 1973.

[4] R.PALAIS, The Principle of Symmetric Criticality, Comm. Math. Physics 69 (1979), 19 - 30 .

[5] A.SANINI, Applicazioni tra variete riemanniane con energia critica rispetto a deformazioni di metriche, Rend. Math. 3 (1983), 53 - 63.

[6] J.VILMS, Totally geodesic maps, Journal Diff. Geometry 4 (1970), 73 - 79.

[7] C.M.WOOD, Harmonic sections and Yang-Mills fields, Proc. London Math. Soc. 54 (1987).

[8] C.M.WOOD, Harmonic sections, Einstein's equation, and stress-energy, University of Liverpool Preprint, February 1987.

[9] C.M.WOOD, Yang-Mills foliations, University of Liverpool Preprint, in preparation.

LISTE DES PARTICIPANTS

Conférenciers

BADJI, C. — Sénégal
Département de Mathématiques
Faculté des Sciences
Université de Dakar
Dakar-Fann
Sénégal

CAGNAC, F. — Cameroun/France
Département de Mathématiques
Faculté des Sciences
Université de Yaoundé
B.P. 812
Yaoundé
Rép. du Cameroun

DABROWSKI, L. — SISSA/Pologne
SISSA
Trieste

EZIN, J.-P. (Directeur du séminaire) — ICTP/Bénin
(actuelle, jusqu'au 6 septembre 1987)
ICTP
Trieste
(permanente)
Département de Mathématiques
Faculté des Sciences et Techniques
Université Nationale du Bénin
B.P. 526
Cotonou
Rép. Pop. du Bénin

HORVATHY, P. — France/Hongrie
Département de Mathématiques
Faculté des Sciences
Université de Metz
57012 Metz Cedex 1
France

PERCACCI, R. SISSA/Italie
SISSA
Trieste

RIGOLI, M. Italie
ICTP
Trieste

SHABANI, J. Burundi
Département de Mathématiques
Faculté des Sciences
Université du Burundi
B.P. 2700
Bujumbura
Rép. du Burundi

TOURE, S. Côte d'Ivoire
Institut de la Recherche Mathématique
Université Nationale de la Côte d'Ivoire
08 B.P. 2030
Abidjan 08
Côte d'Ivoire

VERJOVSKY, A. ICTP/Mexique
ICTP
Trieste

WOOD, C.M. Grande-Bretagne
Department of Pure Mathematics
University of Liverpool
P.O. Box 147
Liverpool L69 3BX
Grande-Bretagne

WOUAFO KAMGA, J. Cameroun
Département de Mathématiques
Faculté des Sciences
Université de Yaoundé
B.P. 812
Yaoundé
Rép. du Cameroun

Participants

BHATTARAI, H.N. — ICTP/Népal
(actuelle, jusqu'au 31 juillet 1987)
ICTP
Trieste
(permanente)
Department of Mathematics
Tribhuvan University
Kirtipur
Kathmandu
Népal

BUDINICH, P. — Italie
SISSA
Trieste

DUONG, Minh Duc — ICTP/Viet-Nam
(actuelle, jusqu'au 11 décembre 1987)
ICTP
Trieste
(permanente)
Department of Mathematics
Dai Hoc Tong Hop
227 Nguyen van Cu
Hochiminh City
Viet-Nam

GAMEDZE, B.T. — ICTP/Swaziland
(actuelle, jusqu'au 29 septembre 1987)
ICTP
Trieste

HONG, Jiaxing — ICTP/Chine
(actuelle, jusqu'au 27 juillet 1987)
ICTP
Trieste
(permanente)
Department of Mathematics
Fudan University
Shanghai
Chine

KONDERAK, J. ICTP/Pologne
(actuelle, jusqu'au 23 janvier 1988)
ICTP
Trieste
(permanente)
Institute of Mathematics
Jagellonian University
Ul. Reymonta, 4
30-059 Krakow
Pologne

LANDI, G. Italie
SISSA
Trieste

MACHADO, A.H. ICTP/Portugal
(actuelle, jusqu'au 31 juillet 1987)
ICTP
Trieste
(permanente)
Department of Mathematics
Faculty of Science
University of Lisbon
Rua Ernesto Vasconcalez, Bl. C2
1700 Lisbon
Portugal

NONVIDE, S. Côte d'Ivoire
Institut de la Recherche Mathématique
Université de la Côte d'Ivoire
08 B.P. 2030
Abidjan 08
Cote d'Ivoire

NOOR, M. ICTP/Pakistan
(actuelle, jusqu'au 7 décembre 1987)
ICTP
Trieste
(permanente)
Department of Mathematics
Quaid-i-Azam University
Islamabad
Pakistan

SINICCO, I. Italie/Brésil
Dipartimento di Fisica
Università degli Studi di Trieste
Trieste

TRIBUZY, R. ICTP/Brésil
(actuelle, jusqu'au 12 août 1987)
ICTP
Trieste
(permanente)
Departamento de Matematica - ICE
Universidade do Amazonas
Campus Universitario
69.000 Manaus - Amazonas
Brésil

VERA, E. ICTP/Pérou
(actuelle, jusqu'au 17 janvier1988)
ICTP
Trieste
(permanente)
Facultad de Ciencias Matematicas
Universidad Nacional Mayor de San Marcos
Lima
Pérou